D0537264

NATIONAL
GEOGRAPHIC

MIND

NATIONAL GEOGRAPHIC

MIND

A SCIENTIFIC GUIDE TO

Who You Are, How You Got That Way, and How to Make the Most of It

PATRICIA DANIELS

FOREWORD BY TODD B. KASHDAN

NATIONAL GEOGRAPHIC

WASHINGTON, D.C.

CONTENTS

Previous pages: Each person is unique. Opposite: We can harness our drives and abilities to reach our goals.

FOREWORD

EXPLORING THE MIND

BY TODD B. KASHDAN, PH.D.,
Professor of psychology and senior scientist,
Center for the Advancement of Well-Being at George Mason University

Each of us holds a vested interest in how the human mind operates. An understanding of what the mind is, how it works, and how it shapes our personalities and interactions with the outside world is essential to living fully and improving our quality of life. What you hold in your hands is an authoritative guide to the latest information on what psychologists and other scientists know about the mind. We intend for this book to shed light on the mind, so often shrouded in mystery. We hope it will give you ways to learn about yourself, too—how in some ways you are like all people, in other ways you are like some people, and, finally, in important ways you are unlike any person who ever existed.

There is no single strategy for understanding how the human mind defines our personalities and influences everyday decisions. That's why it is essential to explore different domains of knowledge to get the big picture, as this book is meant to convey. Some sections explore the biological systems that provide the organic building blocks for how you feel, think, and behave, including the insights offered by an evolutionary perspective. You will discover how creative people have more sexual success, and how racial prejudice is influenced by facial expressions—examples of deep-seated animal instincts that still make a difference today. There is now widespread recognition that biological and evolutionary principles provide tools for explaining the origins of some of the most complex psychological phenomena, from virtue to evil.

But that doesn't tell the full story. In the long-running battle over whether our personality is dictated by genes or the environment, scientists have provided sufficient evidence to suggest that usually both are involved. Our minds draw upon ancient biological impulses, but they also affect, and are affected by, cultural and social contexts. In the United States, when people are asked to complete the sentence, "I am ________," they usually come up with a personal attribute, an adjective such as "energetic" or "curious." In East Asian cultures such as China, Japan, and Korea, people tend to view themselves as fundamentally connected to others; when asked to complete the same sentence, they are more likely to reference a social role such as "a son" or "a student." How we react to pain, what emotions are desirable and undesirable, and the moral foundations that we live by are each shaped by our fundamental psychological need to form lasting, meaningful connections with other people.

To understand the mind, we must challenge conventional wisdom when appropriate. Traditional metrics of intelligence barely account for what leads to success in childhood and adulthood. An entire field within psychology is dedicated to learning what is most pertinent to cultivating happiness, the capacity to love and be loved, a sense of meaning and purpose in life, and healthy communities and societies. Reward and punishment are insufficient for persuading or influencing another person. Thankfully, this new breed of researchers is giving us insight into the types of motivation that truly facilitate peak performance, social development, and well-being.

These new ideas serve as the foundation for several parts of this book. You will learn about the latest science on sex, murder, sleep, strength, psychopathology, human relationships, and the meaning of life. Considering the breadth and complexity of the human mind, this only seems fair. May this book represent a starting point as you enjoy the pleasures of scientific research on what matters most.

Our sense of our own identities is guided by biology and environment.

INTRODUCTION

SHAPING THE SELF

In 1954, Manford Kuhn and Thomas McPartland published the quintessential psychology quiz, called the Twenty Statements Test. On a sheet of paper, the phrase "I am ______" is repeated 20 times. The test-taker fills in the blanks.

The researchers found that people tend to describe themselves in four ways: as physical beings ("five feet tall"); as social beings ("a daughter"); as self-reflective individuals ("shy"), and as a part of a larger experience ("connected to nature"). In other words, ask "Who am I?" and the constellation of answers adds up to one unique identity—one mind—one self.

When it came to studying the self, psychologists used to divide into camps: mind versus body, nature versus nurture. In recent years, they have realized that the self is a complex mix of traits, instincts, drives, and hidden motivations. New techniques have blurred the distinction between physical and mental: Both inner and outer forces shape us, some built into our genes and some learned.

When we study the self, we discover a mix of traits, instincts, and motivations.

In a way, the mind has its own architecture. At its foundation is biology. Evolution has fostered drives and behaviors that help us survive and reproduce. Fears, loves, and social desires are encoded in our genes. To a surprising extent, these instincts are also built into our neural circuitry. Reactions and emotions, aversions and attractions, can be traced to particular regions of the brain. Stages of growth, from infancy to old age, take everyone on a similar journey as intelligence, language, and emotions develop.

Resting on this foundation are the lower floors of our mental building, those devoted to interacting with the outside world. People are profoundly social animals. Our fundamental need to belong drives our behavior in relationships and in social groups, guided by decision-making processes that may lie below the level of consciousness, and by drives and motives that impel us toward one goal or pull us back from another. Our levels of self-control and grit keep us moving forward, or stalling out, in life. Personality traits, inherited and learned, shape our actions as well.

The upper floors of the mind's edifice are those that consciously or unconsciously control the others. They include the anxieties and distorted thoughts that trouble our lives, as well as the values, strengths, therapies, and interventions that allow us to build a stronger self. No matter how deeply rooted some of our troubles may be, the resilient mind has a host of ways to overcome them.

This book will give you a look at some surprising new findings about why you act the way you do. Along the way, you'll encounter quizzes, exercises, and self-surveys that may help you understand yourself a little better. With these insights, you can start down the road to a more satisfying, fulfilling life.

The MIND &

» PART ONE

BODY

To begin to know our minds, we need to know our bodies. In particular, we need to know our brains, because the mind is the brain in action. Our needs, our motives, our fears and desires arise from deep within the folds of our most complicated organ. They were placed there by millions of years of evolution, and in some cases are a poor fit for the modern world. In these first three chapters, we'll see what we've learned about the biological bases of the mind, and how all humans are alike—and where they differ—in their mental and emotional growth.

» CHAPTER ONE «

THE SCIENCE OF THE MIND

In 2005, a 23-year-old woman survived a traffic accident with severe brain damage. Months later, she remained unresponsive. Her breathing was normal, she slept and woke, but in waking showed no awareness of herself or her surroundings.

This lack of responsiveness earned her one of the saddest diagnoses in the medical lexicon: She was in a permanent vegetative state.

This diagnosis implies that the patient no longer has a mind as most of us would recognize it, a self that thinks and reacts, that possesses memory and awareness. To test whether that was true in this case, British neuroscientist Adrian Owen and colleagues connected the woman to a functional magnetic resonance imaging machine (fMRI), which tracks blood flow within the brain. (Greater blood flow points to areas of increased brain activity.) Talking to her unresponsive body, they asked her to imagine two scenarios: that she was playing tennis, and that she was walking through all the rooms of her house, starting at the front door. The scientists then compared her fMRI scans to those of 12 healthy volunteers asked to imagine the same activities.

The results: The same areas of the brain were activated in the patient and in the volunteers. When she was asked to envision playing tennis, the woman's supplementary motor area, which controls movement, lit up. Her parahippocampal gyrus, a section of the brain used for spatial navigation, became active when she was asked to mentally tour her house. Clearly, she was conscious, aware of her surroundings, and able to think and respond. Her mind was still working inside a silent body.

Owen later used similar methods to study 54 vegetative and minimally conscious patients. The tennis imagery stood in for a "yes" response and the navigation imagery for "no." Five of the patients were able to respond intelligently to his questions. His study and others have

Mind, brain, body: How are they different? Where do they meet?

shown that almost 20 percent of vegetative patients are actually conscious and aware, at least for part of the time.

Owen's 21st-century investigation raised some ancient questions: What is the mind? Where is it? How is it connected to the body, and can it exist without a physical form? Is the mind the same as the self? The mind-body duality issue goes back to the ancient Greeks. Plato, for instance, believed that the soul used the body as an instrument of perception, but "returning into herself she reflects; then she passes into the realm of purity, and eternity, and immortality, and unchangeableness." Whether you realize it or not, you are weighing in on the duality side of the mind-body divide when you say "my body" or "my brain"; the construction assumes that you are an entity possessing a separate physical body.

The mind is the brain in action.

Today, few psychologists make that distinction. Owen's research and a wide range of other studies tell us that the mind is the brain in action. Without a brain, we have no self. Even our most evanescent longings, our most spiritual moments, can be traced to our neural circuitry. And that neural circuitry, marvelously intricate, has developed over millions of years of evolution.

“We do not see things as they are, we see things how we are.”

AUTHOR ANAÏS NIN

Our behavior is affected by a host of outside influences, and can certainly be consciously modified, but first and foremost we are creatures controlled by an ancient brain. To begin to understand ourselves, we have to understand our biological history.

THE FIRST PSYCHOLOGISTS

The scientific study of the mind actually has its roots in biology—specifically, in the groundbreaking discoveries of the 19th century. Standing tall over the era is the figure of Charles Darwin. With *On the Origin of Species* (1859) and *The Descent of Man* (1871), Darwin not only laid out the rules of evolution through natural selection, but also empowered scientists to view human beings as members of the animal kingdom, organisms whose bodies and brains can be explained in a purely biological way. Doctors had already begun to study the transmission of impulses along the nerves and to map the location of certain functions within the brain. In 1861, for instance, French surgeon Paul Broca encountered a patient who had lost his verbal abilities and could say nothing but "tan"—in fact, the unfortunate man became known in the hospital by that name. After Tan's death, Broca conducted an autopsy and discovered a lesion in the brain's left hemisphere. He concluded (correctly) that the region, now known as Broca's area, was in charge of speech production.

Many physiologists of the time believed that mental processes could be best understood by studying the brain as an organ that reacts to and learns from sensations. German scientist Wilhelm Wundt, often considered to be the founder of modern psychology, established a laboratory at the University of Leipzig in the 1880s in which he and students from around the world conducted experiments on sensory perception and learning. Wundt scorned the use of introspection to study mental processes, believing that asking an

experimental subject what he had thought or felt during an experiment was useless and unscientific.

American philosopher and psychologist William James was one of the many scientists who visited Wundt's lab. Although he appreciated a rigorous approach, he was never convinced that psychology could yet be called a science. James instituted the first psychology course taught in America at Harvard and wrote a hefty tome on the subject, *The Principles of Psychology,* in 1890. Nevertheless, he characterized the field as "a string of raw facts; a little gossip and wrangle about opinions . . . but not a single law in the sense in which physics shows us laws, not a single proposition from which any consequence can causally be deduced. This is no science, it is only the hope of a science." James was one of the first psychologists to suggest that scientists apply evolutionary theory to the study of everyday behavior. "Why do we smile, when pleased, and not scowl?" he wrote. "Why are we unable to talk to a crowd as we talk to a single friend? Why does a particular maiden turn our wits so

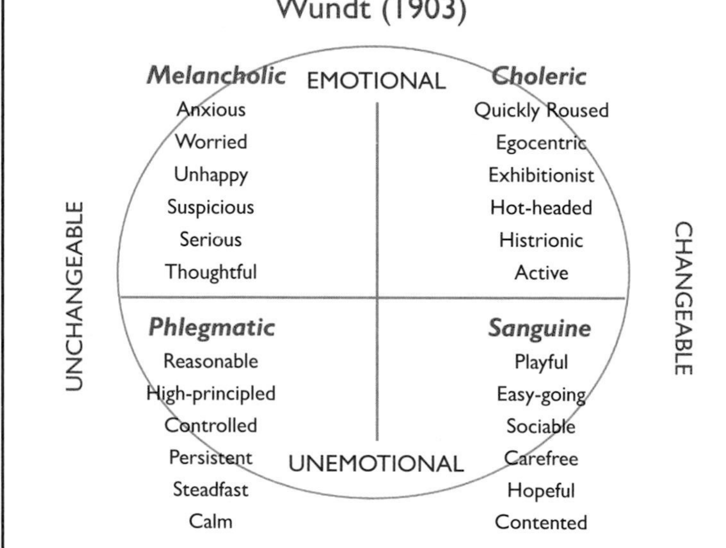

Early American psychologist William James suggested scientists apply the new science of evolution to study behaviors like smiling.

upside-down? The common man can only say, *Of course* we smile, *of course* our heart palpitates at the sight of the crowd, *of course* we love the maiden, that beautiful soul clad in that perfect form, so palpably and flagrantly made for all eternity to be loved! And so, probably, does each animal feel about the particular things it tends to do in the presence of particular objects." On the other hand, he also believed in the value of introspection, in "looking into our own minds and reporting what we there discover." Among his contributions to the field was the phrase "stream of consciousness."

Always open-minded, James was one of the first Americans to acknowledge the work of Sigmund Freud, whom he met when Freud visited Clark University in Massachusetts in 1909. Freud would bring a completely different perspective to the study of the mind, immensely controversial and immensely influential.

» The Freudian Revolution

Sigmund Freud died in 1939. In

Freud's ACTUAL Couch

the decades since his death, psychologists have disputed or disproved many of his theories and therapeutic approaches. And yet Freud remains one of the most influential thinkers in the history of science. His ideas about the unconscious mind, instincts, repressions, defense mechanisms, and more have entered the popular vocabulary and still shape our judgments about ourselves and others. Our vision of therapy is typically still the Freudian one, with the patient supine on a couch and the (bearded, Viennese-accented) therapist taking notes nearby.

Trained as a neurologist, the young Freud began to develop his ideas about the unconscious mind as he treated patients with hysteria—a catch-all diagnosis, usually applied to women, describing emotional extremes and psychosomatic symptoms such as weakness or paralysis. He based his treatment upon a therapy described to him by his mentor, physician Josef Breuer. Breuer's patient Anna O. had come to him with hallucinations, partial paralysis, an inability to drink liquids, and other disabling problems. Breuer treated her by guiding her through her memories. For instance, in therapy she recalled seeing a dog drink from a glass of water, an act that disgusted her at the time, but that she had forgotten. After describing the memory, Anna O. regained her ability to drink. She also reported to Breuer a disturbing dream she once had in which she was unable to fend off a snake with her suddenly useless arm; after recounting the dream, she regained use of her formerly paralyzed arm.

It was Anna O. who dubbed the therapy "the talking cure." She was not, in fact, cured by the method, although her symptoms did abate temporarily. Only a later stay in a sanatorium brought her permanent relief. Nevertheless, the link between

"The interpretation of dreams is the royal road ... to the unconscious activities of the mind."

PSYCHOANALYST SIGMUND FREUD

hidden memories and illness was convincing to both Breuer and Freud, and Freud began to treat his patients with the technique. As he worked with his patients, he developed his fundamental methods of analysis, which came to include the use of free association and dream interpretation to retrieve repressed memories and uncover emotional conflicts.

Freud was not shy about analyzing himself, and his own dreams formed a key part of his first book, *The Interpretation of Dreams,* published in 1900. Calling dreams "the royal road to the unconscious," Freud distinguished between a dream's manifest content—the actual events of the dream—and its latent content—the underlying meaning

Freud established traditions in therapy, terminology, and dream interpretation.

SIGMUND FREUD

Born Sigismund Freud in Freiburg, Moravia (now the Czech Republic), the future psychoanalyst was launched into an outsider's life. Freud was a Jew in an anti-Semitic culture, an atheist in Jewish circles . . . and an intellectual leader who broke with almost all of his followers.

The son of a small-time merchant in Freiburg, young Sigmund was a brilliant and hardworking student. He spoke eight languages and entered the University of Vienna at the age of 17, where he embarked upon a degree in medicine. Early in his career Freud specialized in neurology and brain anatomy, but by 1886 he switched to the field that truly fascinated him, nervous disorders.

Freud's fame grew in the 20th century. He accumulated a distinguished group of followers who called themselves the Wednesday Psychological Society. The group eventually included Alfred Adler, Otto Rank, and Carl Jung—all of whom developed divergent theories and broke away from Freud in the early decades of the century.

Freud left Austria in 1938, after the Nazis invaded. By that time he had long been suffering from cancer of the jaw, the legacy of his heavy cigar-smoking habit. He died in London from a deliberate overdose of morphine in 1939.

of the dream. Later publications, including *The Psychopathology of Everyday Life*, spread Freud's ideas around the world.

Those ideas were radical. Freud saw the mind as a complex energy system, powered by the instincts of self-preservation and sexuality. Most of the mind's workings, in his conception, were unconscious, consisting of memories, emotions, and thoughts too disturbing to be consciously acknowledged. Many of these memories dated to the aggressive, sexualized days of early childhood, in which the child moved through oral, anal, and other stages to reach a mature, genital stage. According to Freud, childhood was dominated by the id, the mind's primitive, pleasure-seeking part. Later, the ego developed to restrain the id,

FOCUS

CARL JUNG

Swiss psychotherapist Carl Jung was the most famous of Sigmund Freud's followers, and he was probably the most influential among those who broke from Freud. The son of a rural pastor, Jung was raised in the beautiful countryside of Laufen, Switzerland. He was a lonely boy—none of his siblings survived—and was troubled by the eccentric behavior of his depressed mother, who told him that spirits visited her at night. "The feeling I associated with 'woman' was for a long time that of innate unreliability," Jung later wrote. "'Father,' on the other hand, meant reliability and powerlessness." Jung himself was not a reliable spouse. He married a wealthy Swiss woman, Emma Rauschenbach, in 1903, and had five children with her, but conducted numerous affairs throughout their long married life.

Interested in both spirituality and psychiatry, Jung began to study medicine at the University of Basel. After a correspondence with the increasingly famous Sigmund Freud, Jung became a close friend and associate of the older psychiatrist. By 1913, however, Jung had begun developing his own ideas and broke away from Freud. Interested in dreams and symbols, Jung proposed the existence of two types of unconscious states: the personal unconscious and the collective unconscious, a repository of symbols and archetypes shared by all people. He also pioneered the idea of introverted and extraverted personalities, as well as the therapeutic notion that people must integrate their unconscious and conscious selves in order to become whole.

Jung remained interested in the role of myths and symbols throughout his life and even published a book on the meaning of UFOs in 1959. In 1961, he died at his house in Küsnacht, Switzerland.

followed finally by the superego, which incorporated the strictures of society into the mind.

Critics of Freud's time were repelled by his emphasis on sexuality. Even so, patients recognized themselves in his descriptions of inner conflict and hidden aggressions. Talking therapies yielded undeniable insights. And popular culture has thoroughly incorporated Freud's ideas. Every time we trace a psychological problem back to childhood experience, or refer to another person as "repressed" or "anal," we're referencing Freud.

Freud's psychoanalytic theories spread around the world, though many psychiatrists modified his techniques as the years went by. Increasingly, his methods came under attack by scientists who declared them unverifiable or, indeed, completely wrong or useless. Notions of repressed childhood sexuality and the inherent inferiority of women, for instance, have been thoroughly debunked. (See chapter 9, page 258) Classical psychoanalysis in the Freudian style is little practiced today. On the other hand, modern cognitive science has, in its own way, revived some of Freud's key theories. The existence of an unconscious mind has been demonstrated in experiments. Internal conflicts, anxiety, and ego depletion are the source of much academic study.

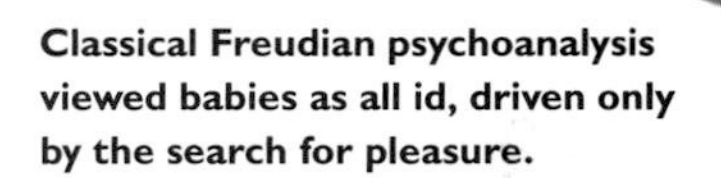

Classical Freudian psychoanalysis viewed babies as all id, driven only by the search for pleasure.

Even the idea of "primitive" instincts, updated, has gained a new life in the field of evolutionary psychology. Darwin's 19th-century science, so influential in early psychology, has once

"Who looks outside, dreams; who looks inside, awakes."

PSYCHOTHERAPIST CARL JUNG

again come to the fore in explaining some fundamental workings of the mind. Darwin predicted as much in *On the Origin of Species.* "In the distant future," he wrote, "I see open fields for far more important researches. Psychology will be based on a new foundation, that of the necessary acquirement of each mental power and capacity by gradation."

THE EVOLVED MIND

Darwin's distant future arrived in the late 20th century. By then, cognitive scientists and neuroscientists were gradually unveiling the mechanics of the brain and its thought processes. Proponents of evolutionary psychology began to consider how natural selection drove the development of those processes and our instinctive behaviors. As cognitive scientist Steven Pinker notes, "cognitive science helps us to understand how a mind is possible and what kind of mind we have. Evolutionary biology helps us to understand why we have the kind of mind we have."

Evolution is the change in the inherited characteristics of a population over many generations. Natural selection is the process that drives those changes. It has two goals: survival and reproduction. Inherited traits that help an organism survive in a particular environment and have healthy offspring will spread through a population over time. Traits that impede survival and reproduction will diminish. These traits can be physical, but they can also be behavioral.

Natural selection works hand in hand with genetic variation, random mutations in the genetic code that create traits that may or may not be beneficial. Take, for instance, red and green beetles. Birds have an easier time seeing, and picking off, red beetles in their green surroundings. Fewer red beetles survive to reproduce, and over time the population of red beetles dwindles and that of green beetles

FOCUS

THE FRIENDLY FOX

Beginning in the 1950s, Siberian geneticist Dmitry Belyaev presided over an unusual long-term experiment. He knew that early humans had managed to turn wolves into dogs. The domestic animals that resulted were not only different physically from their wolf ancestors but were different in behavior as well, becoming loyal, affectionate, and submissive. Belyaev determined to do in 30 generations what our ancestors accomplished in as many centuries: domesticate a wild canid, in his case the silver fox, *Vulpes vulpes.*

Belyaev and his colleagues picked foxes for their project based entirely on docile behavior. Fox pups that allowed experimenters to stroke them, that wagged their tails and whined, were selected for further breeding. After 40 years and 45,000 foxes, Belyaev's Novosibirsk fox farm housed 100 completely domesticated foxes, affectionate and eager for human companionship. "Before our eyes," wrote experimenter Lyudmila Trut, " 'the Beast' has turned into 'Beauty.'" The domestication regime had a fascinating side effect: Although they were chosen based only on behavior, like their dog counterparts some of the foxes now diverged physically from their ancestors. They developed floppy ears and curled tails; in some cases a starlike pigmentation showed up in their fur. Breeding for behavior had evidently changed a complex system of linked genes.

thrives. Red ladybugs, however, can survive in a green environment because of a different adaptation. A bitter taste, which birds learn to avoid, makes their bright color a warning rather than an invitation.

In animals, it's easy to recognize that behavior, as well as physical characteristics, has evolved to promote survival and reproduction. Songbirds instinctively flee from a predator. They sing in the spring to lure mates to their territory. It can be harder to admit evolved behavior in ourselves. Many of us resist the idea that our own choice of a mate or our own peculiar phobia has its roots in prehistory. But there is no reason to think that humans are different from any other organism. Hundreds of thousands of generations have certainly changed our brains from the time of our apelike ancestors. Among other things, we have gained a language. But from the perspective of human lifetimes, evolution is slow. Behaviors that helped our prehistoric human ancestors survive remain embedded in our contemporary brains.

»The Modular Brain

Just as our bodies are modular structures, efficiently compartmentalized into specialized organs, so too are our brains. Different neural circuits are dedicated to solving different problems. Certain parts of the brain govern vision, others memory, sensations, facial recognition, and many other functions. A particular set of pathways is dedicated to sexual attraction, directing our attention to desirable mates. These circuits, or modules, have evolved to use different criteria to solve specific problems. It's a good thing, too. A woman wouldn't want to use the same circuits to choose a chocolate bar and a husband (unless she likes her men sweet and gooey).

These pathways are hardwired into the developing brain, giving us a natural cheat sheet from birth. All animals have separate brain circuits for finding food, selecting a mate, and exploring a habitat. Human

"The ability to be in the present moment is a major component of mental wellness."

PSYCHOLOGIST ABRAHAM MASLOW

babies are born with the ability to distinguish a human face. At less than ten minutes old, a newborn will turn its eyes and head to follow a facelike pattern, but won't pay attention to the same pattern scrambled.

At six to eight months, babies develop a fear of strangers, particularly male strangers, an instinct that illuminates the danger that male animals can pose to other males' children.

Because evolution is slow,

Little Bo Peep is just one of many who experience arachnophobia—the instinctual fear of spiders.

Within minutes after birth, a newborn turns toward a human face.

our inborn behaviors are those that helped us survive once upon a time in our prehistoric environment. Consider fears and phobias. Spiders, snakes, strangers, and heights are among the most common phobias today. Visual tests show that our minds quickly pick out images of snakes or spiders embedded in complex visual arrays, while ignoring other harmless subjects. Yet we rarely have instinctive reactions to much more dangerous modern objects, such as guns or electrical outlets. It made sense for our ancestors to fear and avoid venomous animals or potentially aggressive strangers. It makes less sense today, but we're stuck with evolutionary solutions to outmoded problems.

Our instincts are designed for different ends, and they may compete with one another. Fear of meeting strangers might battle it out with sexual attraction, for instance. We feel this struggle

We helped each other, and we worked it out as a group.
Arranging ourselves in the shape of a heart sounded impossible, but we did it!

within ourselves every time we overcome shyness to approach a new person.

»The Social Environment

Human instincts are also designed for a social world, in which we respond to other people's behavior. Our social environment is as important to us as our physical environment, and possibly even more so. We have evolved as social animals, and research indicates that the presence of others in our lives is healthy and natural. Today, close relationships and rich social networks are linked to better health and well-being. Brain scans show that the brain is calmer and less vigilant when a person has social support. The brain seems to assume that social contact is the norm, the behavioral baseline. Knowingly or unknowingly, we count on having social resources to support us in life.

Social relationships distribute risk. There are obvious advantages to being a member of a group in which some members will alert others to danger. Relationships also save energy. Historically, friends, family, and community have provided help and information and generally reduced the energy an individual would have to expend in the daily labors of life. We instinctively respond to the presence of others with a sensation of support. For instance, participants in a recent study judged a hill to be less steep if they viewed it while standing next to a friend; the longer the friendship, the less steep the hill appeared. When we're faced with tough tasks, in other words, it helps to gather friends around us.

Close relationships, in particular, make life easier and less stressful. Close partners will cooperate to care for youngsters. They'll help when a loved one is sick or injured. Their emotional support provides a mental buffer. In one experiment, women were threatened with an electric shock while either alone,

ASK YOURSELF

SAFETY IN NUMBERS

» How are humans like ostriches?
Both find safety and ease in groups. Researchers who studied wild ostriches saw that they had two competing priorities. In order to spot predators—and these include cheetahs, lions, and African wild dogs—they have to hold their heads up high on their long necks. In order to forage for their usual food, low-growing vegetation, they have to keep their heads close to the ground. Constantly lifting their heads, dropping them, and lifting them again to scan for danger exhausts the lanky birds.

The solution to this problem lies in group vigilance. In a group, only a few ostriches need to be on the lookout, while others graze. The more ostriches, the greater the foraging time. Observers found that the frequency with which an ostrich will raise its head to check for predators varies directly as a function of the number of ostriches in the group. The more ostriches, the less often any one ostrich has to lift its head.

This approach applies to people as well. Being a member of a group gives you the benefits of load sharing, the distribution of labor. It also protects you through risk distribution. Although danger may still threaten a large group, the risk to any one member of the group is small.

» So ask yourself:
When you've been faced with a tough task, whether it's a tricky project at work or a family problem at home, how do you fare when others share the work, as opposed to going it alone? What's your emotional response? Does it ease your anxiety to have others around you? What lessons can you draw from that as you face challenges in your life?

holding a stranger's hand, or holding a partner's hand. Their brains registered the greatest threat-related activity when they were alone, followed by comfort from a stranger, and the least stress when touched by a partner.

Other social instincts have clear evolutionary advantages as well. We are more altruistic toward close relatives than toward distant ones; we are biased toward detecting and punishing cheaters and free riders; we are inherently sensitive to social status and hierarchy and modify our behavior accordingly. All these behaviors keep our families protected and our social environment steady and predictable.

»Gender and Evolution

In 1978, a couple of researchers enlisted male and female college students to wander about their campus and approach people of the opposite sex, saying "I have been noticing you around campus and I find you to be very attractive. Would you go to bed with me tonight?" All the women so propositioned declined. Seventy-five percent of the men agreed, sometimes with comments like "Why do we have to wait until tonight?" The researchers repeated this experiment three more times over the years, including in the 1980s, after the advent of AIDS, with the same results.

Men and women prefer symmetrical looks in the opposite sex

The notion that men want casual sex and women a relationship is a cliché, a stereotype—and a truth validated in many studies worldwide. The idea that men have sex on the brain much more often than women is also borne out by research. For instance, a survey of U.S. 18- to 59-year-olds found that 54 percent of men but only 19 percent of women said they thought about sex every day or several times a day. Other research finds that among dating teenagers, girls typically want a feeling or declaration of love before having sex. Boys, not so much.

These differences in sexual desires and behavior have a clear evolutionary explanation. Both genders are driven to maximize survival and reproduction. However, the costs and benefits associated with these goals are vastly different for men and women. Historically, pregnancy, childbearing, and child-rearing have been hazardous, resource-intensive, lifelong activities for women. Having the support of a committed, strong partner greatly alleviates these risks and costs. Men, on the other hand, bear none of those risks and instead benefit, from an evolutionary standpoint, from spreading their genes as widely as possible. So why do men bother with monogamy? From another, cold-blooded perspective, mating is a marketplace. Sex and fertility can be seen as a valuable resource largely controlled by women. Men "purchase" this resource by courting a woman and investing in a long-term relationship.

Double-standard attitudes toward infidelity are also predictable from an evolutionary standpoint. Around the world and throughout history, women who cheat on their husbands have been judged much more harshly than men who cheat on their wives. Women are giving away a valuable resource, their fertility, to another man and his genes. Men who stray can still impregnate their wives. Their infidelity has no evolutionary cost.

Evolutionary rationales also explain standards of attractiveness. Men and women both prefer symmetrical looks in the opposite sex, an appearance that indicates genetic fitness. Men around the world are drawn to evidence of fertility. They find youthful-looking, narrow-waisted women more enticing. Women are attracted less to looks than to power, preferring affluent men with dominant yet caring personalities.

FOCUS

REGRETS, I'VE HAD A FEW

When it comes to regrets, in most areas men and women feel the same. Both genders are more likely to regret what they didn't do—opportunities they missed, risks they were unwilling to take—than what they did do. When it comes to regrets about romantic relationships, however, men and women diverge. Regarding relationships, women are more likely to regret past actions, such as becoming involved with a jerk. Men are much more likely to regret the road not taken, agreeing more strongly with statements such as "I should have tried harder to sleep with ___." From an evolutionary perspective, this makes sense. Women benefit from a careful choice of partners, while men are motivated to seek out many mates.

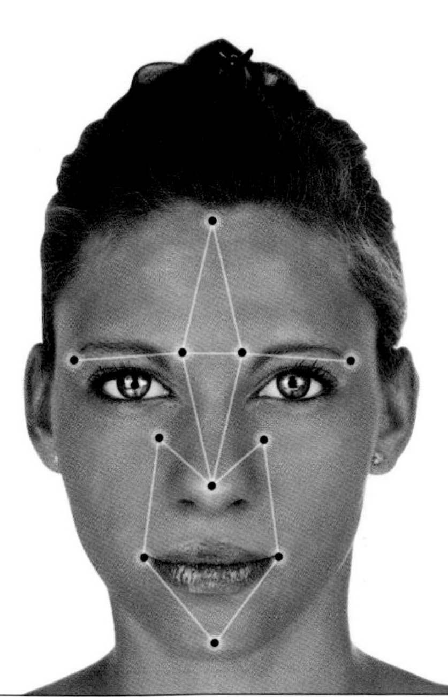

Facial symmetry generally signals genetic fitness to the opposite sex.

“It has been said that man is a rational animal. All my life I have been searching for evidence which could support this.”

PHILOSOPHER BERTRAND RUSSELL

Subtle differences in memory skills between men and women also point to an evolutionary influence. Assuming that, throughout most of human history, men have been hunters and women gatherers, evolution should favor stronger navigation memory in men, but better object memory in women. This turns out to be true, although the difference in abilities is slight. In one study, men and women visited six different food stalls at a farmer's market. The researchers then showed the participants a map of the market grounds and asked them to point out the location of each stall. Women were significantly better able to locate the stalls on the map as well as to remember where to find the food with higher calorie content, skills that would have been useful for a foraging woman in earlier times. This kind of memory is a Stone Age skill living on in a modern skull.

Patterns of partnership reflect our inborn urge to pass on genes to the next generation.

»The Effects of Culture

As satisfying as evolutionary explanations can be, it's vital to keep in mind the fact that instinctive tendencies are only one factor among many in determining human behavior. They can be easily outweighed by the forces of learned behavior. In the West, the standards of sexual attractiveness have varied widely in recent centuries, from the Rubenesque curves of the 17th century to the boyish physique of recent decades. These preferences have changed too quickly to be the effect of evolution. Evolutionary reasoning also says that men should always prefer mates who are virgins, because that would ensure that they, and not competitors, were parents of future children. However, surveys show that preferences for chaste mates vary by culture. Chinese, Indian, and Iranian men, for instance, prefer virgins, while men in Sweden, Germany, and France either don't care or actually find chastity undesirable. Almost all cultures endorse monogamy, and their populations follow suit, with men and women eventually seeking long-term commitments. Cultures with less gender inequality will show fewer differences between men and women in their preferences for mates. Evolution is hardly destiny, as far as behavior goes. "My genes made me do it" is not a valid excuse.

»Competing Motives

Evolution also shapes our motivations. In 1943, American psychologist Abraham Maslow brought a new perspective to his field. Freudian psychologists had been exploring neurosis and dysfunction; behavioral psychologists rejected that focus in favor of a mechanistic approach toward learning and stimulus. Maslow took a third path, examining human behavior in terms of motives. He believed that people are driven toward their goals by innate needs, motives found in every human being. Although he didn't invoke evolutionary theory, his ideas reflected a basic tenet of that science, which says that all behavior is goal-oriented. Natural selection has adapted us to drive toward desired ends, such as survival and reproduction.

Maslow identified five basic needs or motives common to all humans and arranged them in a hierarchy, his now famous pyramid. At the base are fundamental physiological needs, such as food, water, sleep, and sex. Above those are safety needs: shelter, security, predictability. At the next level are motives of love and belongingness, which include all varieties of affection and social connection. The next rung is

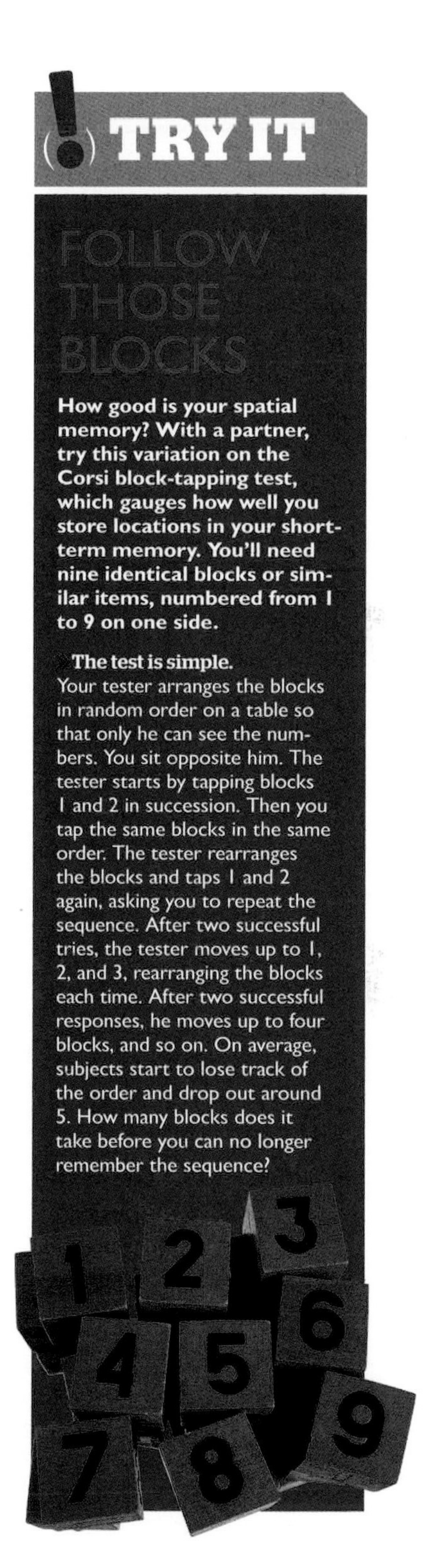

! TRY IT

FOLLOW THOSE BLOCKS

How good is your spatial memory? With a partner, try this variation on the Corsi block-tapping test, which gauges how well you store locations in your short-term memory. You'll need nine identical blocks or similar items, numbered from 1 to 9 on one side.

The test is simple. Your tester arranges the blocks in random order on a table so that only he can see the numbers. You sit opposite him. The tester starts by tapping blocks 1 and 2 in succession. Then you tap the same blocks in the same order. The tester rearranges the blocks and taps 1 and 2 again, asking you to repeat the sequence. After two successful tries, the tester moves up to 1, 2, and 3, rearranging the blocks each time. After two successful responses, he moves up to four blocks, and so on. On average, subjects start to lose track of the order and drop out around 5. How many blocks does it take before you can no longer remember the sequence?

esteem or respect, the need to feel good about ourselves and to receive respect from others. And at the top is self-actualization, the need to achieve one's potential.

Maslow organized the motives into a pyramid to show that, on average, some motives always take priority over others. Primary needs for food, shelter, and safety always trump social or artistic motives. A starving man will put the search for food ahead of self-expression. Maslow also believed that the hierarchy reflected human development, with the most basic needs being those experienced by all infants, with the desire for affection, esteem, or self-actualization arising later and later in the life story.

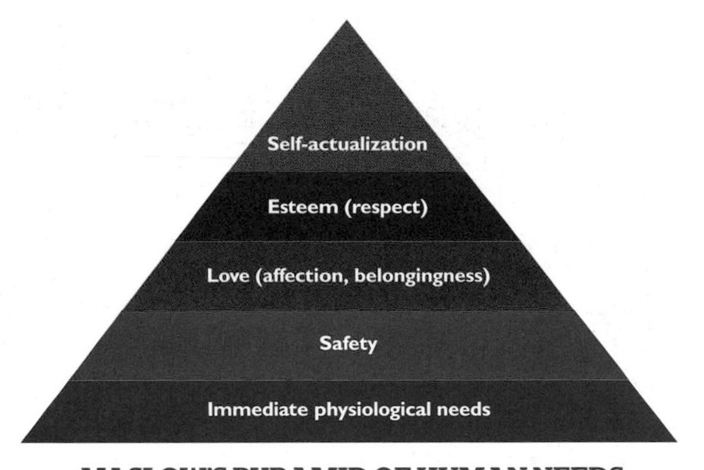

MASLOW'S PYRAMID OF HUMAN NEEDS

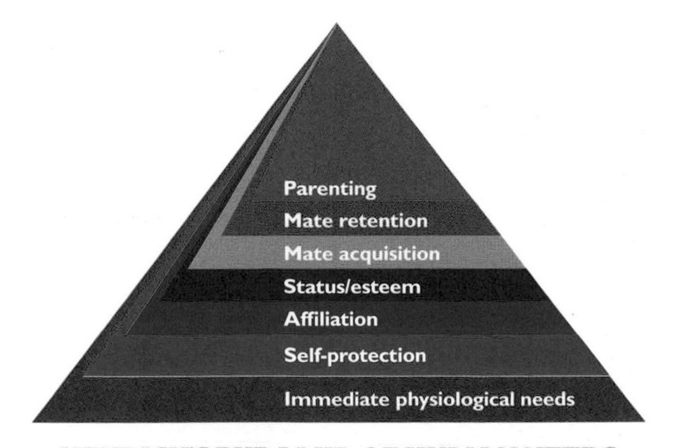

KENRICK'S PYRAMID OF HUMAN NEEDS

Maslow's pyramid was clear and made sense to psychologists and laypeople alike. It helped to shape the field of positive psychology, which focuses on how people can achieve well-being. And to some extent, it has been borne out by research. People do seem to have strong, identifiable motivation systems for basic physiological needs, safety, social connection, and status. Self-actualization as a basic need is more dubious. It's true that people often pursue creativity and self-expression, and have been known, for instance, to continue in jobs that no longer paid them. But empirical support is especially lacking for this layer of the pyramid, as well as for the model as a whole.

Contemporary psychologist Douglas Kenrick and colleagues have proposed a revised pyramid of needs that would more clearly align with evolutionary motives. At its base it resembles Maslow's, but his self-actualization category has been replaced by the needs for mate acquisition, mate retention,

Bonding and status are among the higher human needs in Kenrick's pyramid.

and parenting. Like Maslow's hierarchy, Kenrick's is organized according to priority and child development. Needs such as hunger and self-protection are present from infancy. Social bonding and esteem become gradually more important during childhood, while young adults use the social skills they've acquired to seek out and attract mates, all of which is logically followed by parenting.

The nested architecture of Kenrick's pyramid highlights the idea that needs are overlapping; later, or higher, motives don't replace earlier ones. All our needs remain in place during life, sometimes competing with or overtaking other motives depending upon the situation. We can seek out status and romantic partners at the same time. These motives don't go away, but might recede when we are faced with threats to our safety or basic survival. Our motives mix and come to the fore or fade back depending upon context. If you're having lunch with your boss and trying to impress her, but find a scorpion creeping up your leg, survival outweighs status and you'll jump up and knock it away. If it's an ant on your leg, status now dominates survival and you might let the insect wander.

An ancient environment shaped our bodies and psyches.

»Outmoded Adaptations

Our bodies and our psyches are shaped by the challenges of an ancient environment. Some of those challenges have vanished in modern times, but the adaptations have not. Evolutionary change requires thousands of generations, so we're coping with new foods, technologies, and culture using bodies and brains designed for a hunter-gatherer lifestyle.

Obesity and diabetes are physical symptoms of this evolutionary mismatch. Sugars and fats, with their high energy content, were useful but hard to come by in early human history. Our bodies evolved to crave these foods, seek them out, and store the excess calories against times of famine. Now, these nutrients are all too plentiful and famine, at least in the developed world, is rare, but our bodies haven't heard the news. We're still gobbling up sweets and storing the excess calories—on top of other excess calories. The next time you reach for that brownie, you might reflect that your ancestors might have needed it, but you don't.

Phobias, as mentioned earlier, are another clear example of outdated adaptations. Snakes, spiders, and wide-open spaces don't pose the threat they once did, but tell that to your evolved brain. Even some phobias tied to the modern world can be explained by ancient fears. Fear of flying can't have evolved since the invention of airplanes. However, fear of extreme heights, fear of being trapped in enclosed spaces, and fear of being out of control in an unfamiliar situation would have been adaptive in earlier times. Taken together, they make the experience of airplane travel instinctively terrifying.

Separation anxiety, too, once played a larger role in survival. It made sense for early human children to fear being taken away from their caregivers, upon whom they were entirely dependent. Today, children might venture farther from the home to preschool or day care, but built-in separation anxiety can make their lives (and their parents' lives) unnecessarily difficult. Understanding that, and knowing that reassuring experiences will overcome their fears, can help

A child's separation anxiety harks back to earlier times, when distance from a parent could mean danger.

parents and caregivers weather the transition.

Modern medicine has made some instincts not only irrelevant, but also potentially dangerous. In the pre-contraception age, male sexual jealousy helped to ward off other men and ensure that a man's mate bore his own children. With contraception, though, that motive should be erased in any rational man. Similarly, in societies where women can support themselves and their own children, emotional jealousy should be irrelevant. But of course, not nearly enough time has passed to erase these maladaptive reactions from our brains. In any event, if contraception had existed in early history, it might have been viewed as a threat to reproduction. As Stephen Pinker notes, "Had the Pleistocene savanna contained trees bearing birth-control pills, we might have evolved to find them as terrifying as a venomous spider."

Phobias are clear examples of outdated adaptations.

»Hindsight and Unanswered Questions

Evolutionary psychology is a work in progress, a field that still has its critics and conundrums. At its sloppiest, it is the art of hindsight. It's all too easy to take a phenomenon and reason backward from it to the evolutionary

“We cannot become what we need to be by remaining what we are.”

BUSINESSMAN AND WRITER MAX DE PREE

explanation of your choice. People are altruistic? That perpetuates their genes in a secure, cooperative society. People are homicidal? That perpetuates their genes at the expense of the less fit. This sort of easy explanation is more common in the popular press than it is among scientists, who try to limit themselves to testable predictions based on what we know for sure about ancient human societies.

Some behaviors still confound the rules of natural selection. Suicide is a prime example. Nearly 40,000 people commit suicide in the United States each year, a large percentage of them between the ages of 15 and 44. It is hard to propose an evolutionary rationale for youthful suicide, a behavior that surely should have been weeded out from the population long ago. It's possible that suicide is some kind of nonadaptive by-product of another behavior, but a sound explanation has yet to be found.

Despite these unanswered questions, a considerable amount of research has supported the idea that evolution has shaped our brains and, by extension, everything controlled by our brains: our innate, deep-rooted behaviors, our personalities, and even our basic sense of our own identity. We are who we are because our ancestors faced particular challenges in their environments and overcame them with mental adaptations. Now, our brains incorporate those adaptations to make us modern humans—happy, fearful, adventurous, loving, and curious about our own selves.

Many functions of the human mind can be traced back to the pressures of primitive survival.

» CHAPTER TWO «

THE MIND IS WHAT THE BRAIN DOES

Do we know our own minds? Studies of the brain seem to show that we don't. ¶ In rare operations known as hemispherectomies, communication between the two halves of the brain is impaired, usually because surgeons have cut a connecting nerve bundle (the corpus callosum) in a last-ditch effort to stop severe epileptic seizures. In general, people can function fairly well without that left-right connection. They retain their previous personalities and intellectual abilities. However, the operation highlights the fact that the two hemispheres can operate almost independently of each other, as if the patient has two minds that are sometimes in conflict.

In a classic experiment, researchers sat split-brain patients in front of a screen and asked them to stare at a dot in the middle. Then they flashed the word "HEART" so that "HE" appeared on the left side of the screen and "ART" on the right. When they asked the patients which word they saw, they replied "ART"—the word that was transmitted via the crisscross visual pathways from the right side of the screen to the left side of the brain, where the main language centers reside. But when experimenters asked the same patients to point with their left hands to the word they had seen, they pointed to "HE"—the word that reached their right hemisphere, which controls the left hand. This response startled the patients, who had no conscious awareness of having seen that part of the word. Other split-brain patients have been able to read two books simultaneously, one with each eye, or draw two different shapes at the same time with both hands. In some patients, the left hand literally did not know what the right hand was doing; one hand would unbutton a shirt while the other hand buttoned it again.

Perhaps most disturbingly for those of us who like to think we're aware of our minds, people with conflicts between the

The Brain

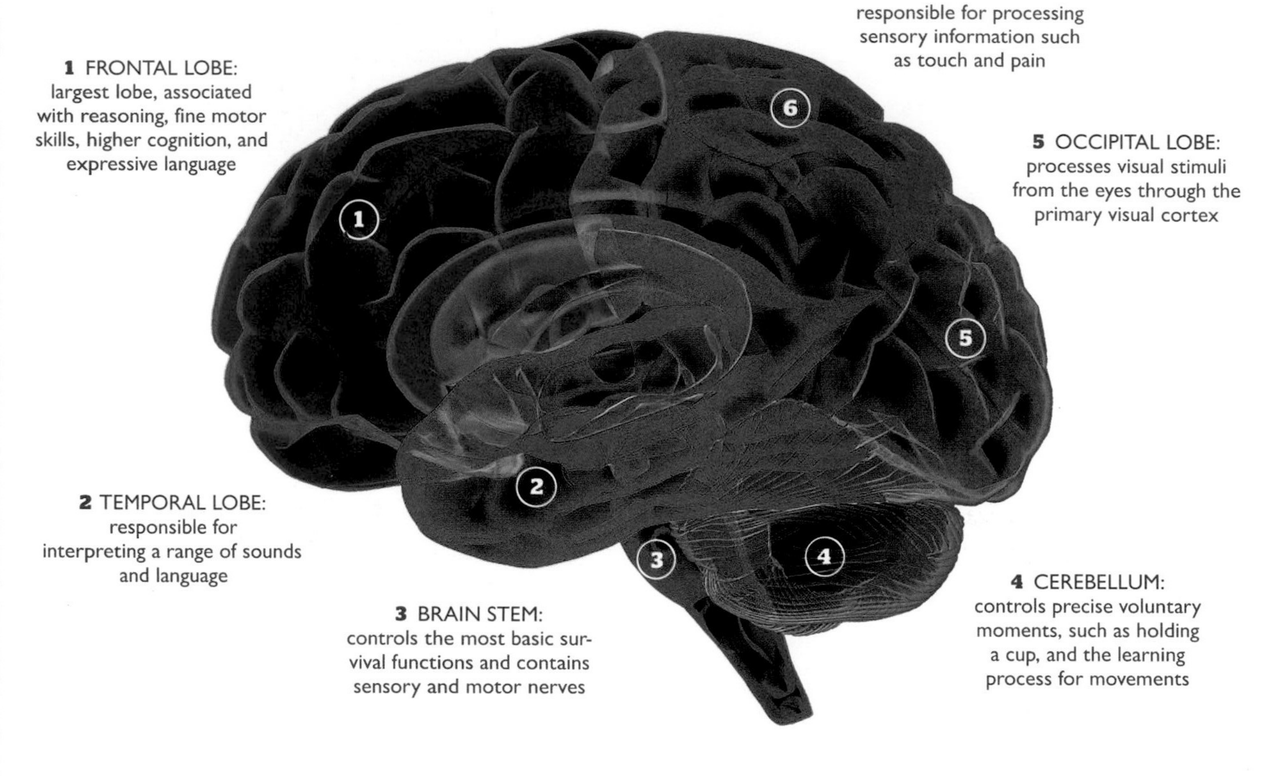

The cerebrum is the largest part of the brain, divided into the left and right hemispheres. The outermost layer of the cerebrum, the cerebral cortex, is divided into four sections called lobes.

two hemispheres can't acknowledge them. A patient whose right hemisphere is given the instruction "walk" will start walking. When asked why he's walking, he doesn't say he was told to walk or that he doesn't know. His rationalizing left hemisphere comes up with a reason. "I'm going into the house to get a Coke."

Do we really control our own behavior? Or do our brains have their own agendas, independent of conscious awareness? Just as evolutionary science has shown that some of our behavior is built into our brains, modern advances in neuroscience have demonstrated that those brains will process information and make decisions essentially on their own. Many psychologists now agree with cognitive scientist Marvin Minsky, who said that "the mind is what the brain does." A wide range of emotions and behaviors can be tracked directly to regions of the brain. Memory, attention, judgment, perception, empathy, emotions, and the recognition of self and others are all rooted in neural processes. Disconcertingly, many of these processes operate under the mental radar, showing up in brain scans even when we are not consciously aware of them. When you find yourself behaving in inexplicable ways, there could be a good reason for that. Your hidden brain made you do it.

TWO MINDS

Our mental abilities can be mapped to particular portions of our cerebrum. Each of the brain's two hemispheres, it has long been known, receives visual information from the opposite eye: The right hemisphere takes in what the left eye sees, and the left hemisphere takes in what the right eye sees. Each hemisphere also controls the actions of the opposite side of the body, with the right brain moving the left side and the left brain moving the right.

The two hemispheres mirror each other, but some mental functions dominate in one half or the other. Mathematics, logic, and retrieval from memory predominate in the left hemisphere; spatial abilities, musical and artistic understanding, intuitive reasoning, and recognition of faces and emotions are housed mainly in the right hemisphere. Language is found in both, but predominates on the left side, which is responsible for rationally interpreting speech. The right side also recognizes most words, but has trouble with syntax. To the right hemisphere, "the flying planes" and "flying the planes" are indistinguishable.

FOCUS

BRAIN BASICS

The pinkish-gray, wobbly, wrinkled organ in our skulls is a miracle of biology. With roughly 90 billion neurons creating immensely complex webs of connection, the brain not only controls the body, but also creates our ineffable sense of consciousness and identity.

At the most basic level, the brain can be divided into three structures: the hindbrain, the midbrain, and the forebrain. The hindbrain and the midbrain mostly handle ancient functions such as breathing, heart rate, and coordinated movement. When we think of the mind, we're usually thinking of the forebrain—the big cerebrum and the little structures buried within it.

The cerebrum is split by a deep groove into right and left hemispheres. In general, the hemispheres mirror each other in structure and functions (although most language processing happens on the left). At the back of each hemisphere is the occipital lobe, which processes vision. Around the side is the temporal lobe, home to sound, speech, and some memory. Toward the top, the parietal lobe handles movement, calculation, and some forms of recognition. In front, behind the forehead, each frontal lobe manages conscious thought, decision-making, and emotional responses, among other things.

Tucked under and inside the cerebrum, in the center of the brain, are the curving little structures that make up the limbic system. Emotions and information that affect our survival flow through these modules. Signals from the senses are sorted and distributed here. Each amygdala detects and communicates fear. The hippocampus lays down our long-term memories, basic to our own identities.

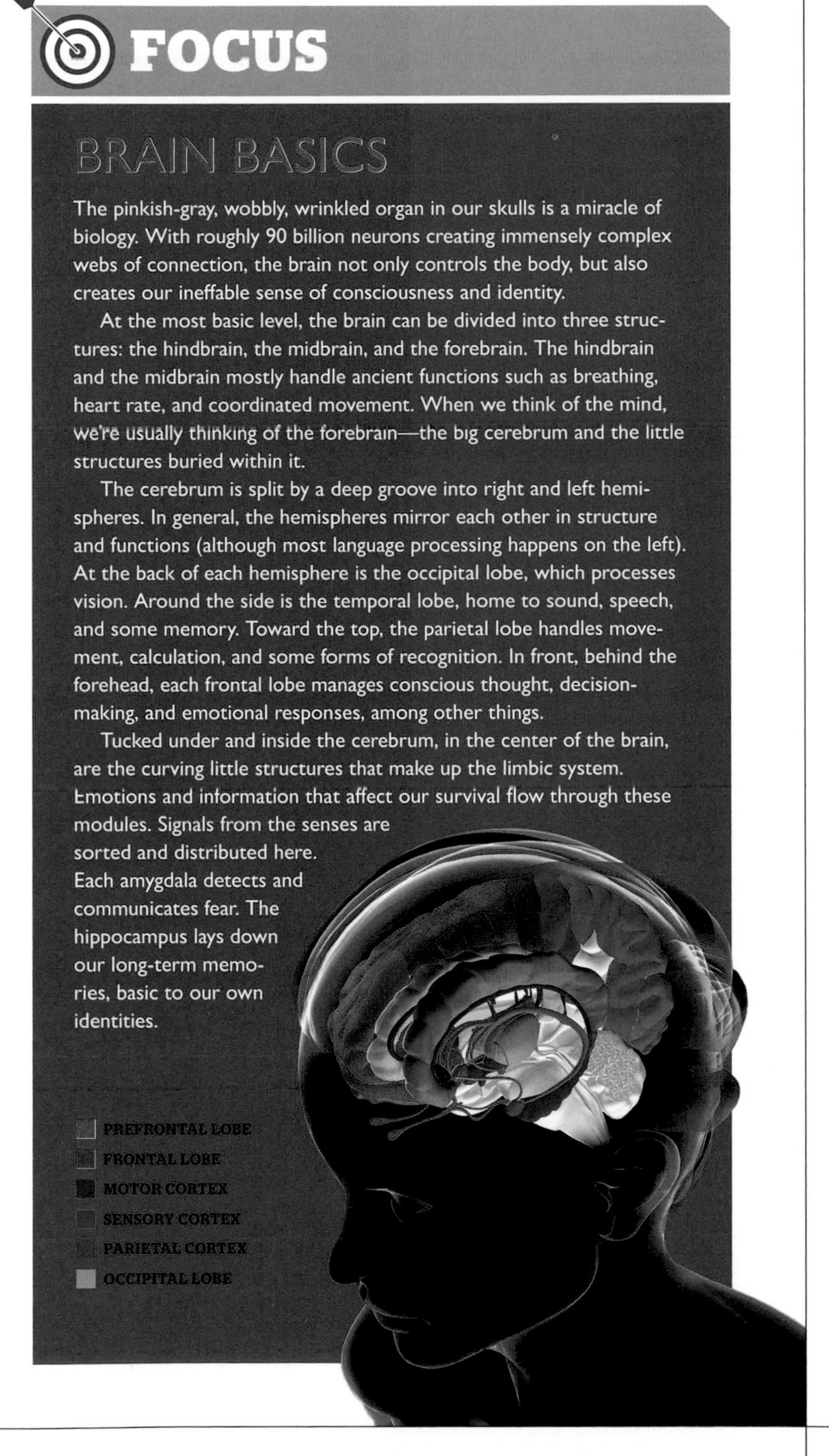

TRY IT

PUT ON A HAPPY FACE

Look squarely at the center of each of the two faces above. Which one looks happier to you? If you're like most people, the face on the right looks more cheerful. This may be because the right hemisphere, which receives information from your left eye, is much better at interpreting emotions in expressions.

Specialization allows the brain to work more efficiently. Communication between the two halves ensures that it works smoothly. The corpus callosum transmits information in a flash, so that we instantly merge the view from both eyes and coordinate the two sides of our body without conscious thought. Knowledge of the literal meaning of a word, born in the left brain, combines with a sense of its meaning in context, taken from the right. To a normal person, the division between the two hemispheres is invisible. Our thoughts and actions seem to stem from one, single self.

»The Mind as Iceberg

Split brains aside, even normal brains have been found to possess two tracks: the conscious processing of information—such as sensations, thoughts, emotions, and decisions—and the unconscious processing of information below the level of awareness. Like the unseen mass of an iceberg, unconscious processing vastly outweighs conscious. According to one estimate, the senses take in 11 million bits of information a second, but we consciously think about only 40 of them. This is not a bad thing. We couldn't function if we were simultaneously aware of every noise, every touch, every breath, every shift in posture, in each moment of our lives.

One example of unconscious awareness is the phenomenon of blindsight, which arises in people with cortical blindness. These people have normal eyes and visual pathways, but have suffered damage in the areas of the brain that govern vision and object recognition. Hold a ball in front of such a person and ask her to identify it, and she'll reply that she can't see it at all. Ask her to reach out and grasp it, however, and she can do so unerringly. Some visual information has reached undamaged parts of the brain that are responsible for registering location, without rising to the level of conscious awareness.

Similar to the unseen mass of an iceberg, the mind processes much more information unconsciously than we notice consciously.

Blindsight has its counterparts in the hidden sides of other mental processes, including memory, decision-making, prejudices, social interactions, emotional responses, and more. Much of modern psychological research is devoted

Conscious
(above water)
Unconscious
(below water)

to understanding these undercover motivations and how they affect our behavior.

ALTERED STATES

The hidden mind is nowhere more evident than in the altered states of consciousness that are built into our daily lives. Sleep, occupying roughly eight out of every twenty-four hours in our lives, takes the mind into a dark zone of shifting mental patterns and hallucinatory images.

The mind's activity during the sleep cycle is different from that during waking.

While slumbering, our awareness of external sensations is dimmed but not completely extinguished: The familiar sound of an air conditioner turning on and off may not disturb us, but the scrape of a footstep in a downstairs room will shock us awake.

Sleep is not the same as unconsciousness, but rather its own unique phenomenon. Electroencephalographs hooked up to sleeping brains reveal that we cycle repeatedly through five stages of sleep during the night. Stage one, marked by irregular brain waves, passes into stage two, a deeper and more relaxed period in which the brain shows occasional spikes of activity. The brief, transitional stage three moves into the deepest part of sleep, stage four, in which the brain emits slow delta waves.

From this deepest well of sleep, the brain climbs out and becomes more active until it enters REM, or dreaming sleep. Heart rate quickens and the eyes flick back and forth beneath closed lids in the rapid eye movement that gives this stage its name. More than 80 percent of people report dreams if wakened

STAGES OF SLEEP

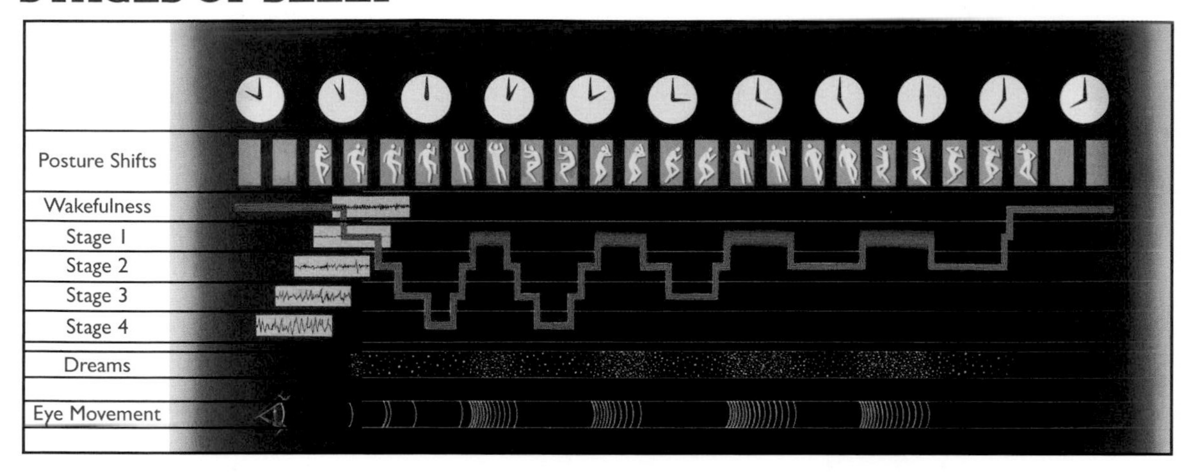

during this stage. Meanwhile, the brainstem shuts down large muscle movements so that, aside from twitching eyes or fingertips, the body is effectively paralyzed. This prevents it from acting out the movements in its dreams. Sometimes this paralysis lasts into the transition to wakefulness, creating an eerie threshold state in which the sleeper may experience hallucinations as well as the unsettling sensation of being unable to move.

The reason for this dramatic shift in consciousness every day? No one knows. Sleep remains one of the great scientific mysteries. From an evolutionary perspective, sleep seems both wasteful and dangerous. Sleeping animals don't reproduce, and they could be vulnerable to predators. And yet sleep also seems to be essential, experienced not only by humans but also by all birds and mammals. Without it, we perish. People with a rare inherited form of insomnia typically die within a few years. Rats kept from sleep also die within weeks, though their autopsies reveal no physical cause.

Many theories have been advanced, but none proven, about the reason for sleep. Some scientists believe it must bolster brain functioning. Experiments have shown that during sleep the brain may consolidate recently learned information, while weeding out other, underused connections. Other researchers think sleep serves the body, conserving energy and resources while allowing for rapid arousal if danger threatens. So far, we know nothing for sure except that sleep is essential.

Without sleep, we perish.

» Anesthesia

Drugs also alter consciousness, and few effect a more dramatic change than general anesthetics, which induce unconsciousness and block awareness of pain. Just how these drugs do this is not well understood, despite more than 150 years of use. In an unknown way, they block transmission of some nerve impulses in the brain, while allowing for the basics of respiration and heartbeat. Most people awaken from general anesthesia with no memories of the experience. An unfortunate few, however, give the lie to the idea that an inert patient lacks awareness.

In perhaps one or two cases in a thousand, patients report feeling some sensation or hearing doctors talk during an operation. In most cases they don't feel pain, but even so the memories can be traumatic. Many hospitals employ a bispectral index (BIS) monitor, which observes brain waves to measure a patient's level of awareness before surgery. However, recent studies show that the monitor may not prevent awareness under anesthesia. Another approach may shed some light on the problem. Positron-emission tomography (PET) scans of brains recovering from anesthesia reveal that awareness emerges first in old, deep-brain structures such as the thalamus and limbic system. This may explain why devices that monitor just the brain's cortex are less effective.

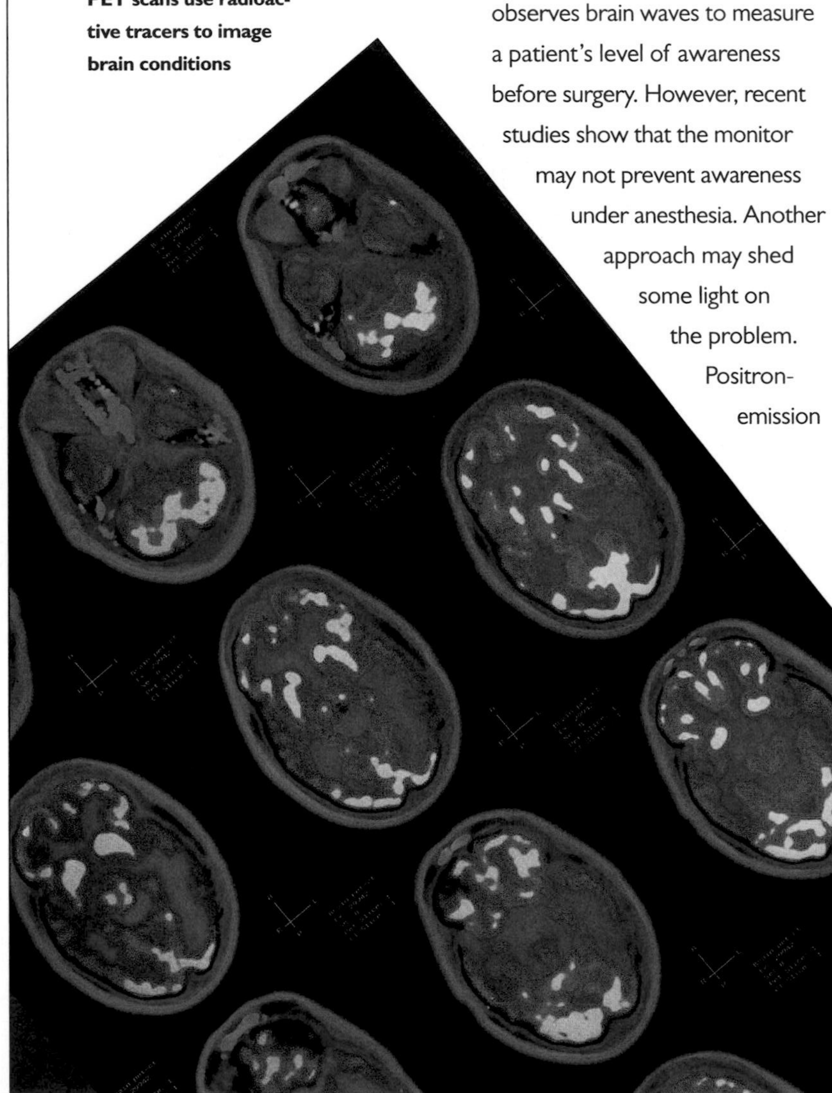

PET scans use radioactive tracers to image brain conditions

»Hypnosis

Like drugs, hypnosis is supposed to induce a change of mental state. The practice of hypnosis—in which a hypnotist induces a trancelike state in the subject—has been around since Viennese physician Franz Mesmer treated his patients using "animal magnetism" in the 1700s. For a time Sigmund Freud employed it in psychotherapy, as did many of his followers. It is still used today to treat a variety of conditions ranging from chronic pain to cigarette addiction. Yet many psychologists believe that the hypnotic experience is less a state of altered consciousness than an example of the intense suggestibility of human nature.

There is no doubt that hypnotic subjects will genuinely respond to irrational commands. After they are brought into a state of relaxation and inward consciousness by the soothing voice of the hypnotist, subjects told that ammonia smells like perfume will sniff it with pleasure. Instructed to forget their actions during hypnosis, they will be genuinely amnesiac. PET scans confirm that people exhibit different brain patterns during hypnosis than they do during everyday awareness, as the regions responsible for mental imagery of actions and sensations become more active. Moreover, hypnotic suggestions from clinical hypnotherapists have certainly helped some people lose weight and block pain. However, the notion that hypnosis is a unique trance state has been replaced by the equally strange finding that people can easily alter their mental states if asked to do so by an authority figure.

Studies from the 1990s show that most of the population is suggestible in this way, to some degree. Almost anyone, told that his eyelids are getting heavier and heavier, will feel them drooping. However, hypnosis per se may have nothing to do with the average person's willingness to comply with odd commands. In some unsettling laboratory experiments in the 1950s, two groups of people—those who had undergone hypnotic suggestion and those who had not—were asked by researchers to perform various antisocial acts, such as

"The mind is what the brain does."

COGNITIVE SCIENTIST MARVIN MINSKY

Facial expressions convey the same emotions in every culture.

The ability to delve into yourself and imagine a different state is the precursor to hypnosis, and most of us have it to some extent. For instance, try to stand straight up and remain immobile while someone tells you you're swaying back and forth, back and forth. The suggestion is difficult to resist.

The true limits of hypnosis become apparent in some of its practitioners' more extreme claims. Hypnotic subjects who are told to regress to a young age will act childishly, but in ways

throwing what they were told was acid into an experimenter's face. The hypnotized subjects threw the acid, but so did those who were not hypnotized. When they were asked about it afterward, the unhypnotized group told the researchers that they assumed the action would be safe, because responsible authority figures were in charge. (Notably, when the experimenters informally asked their own colleagues to perform the same actions, all refused.)

The people who are most apt to become hypnotized are those who have faith in hypnosis and believe that they could be hypnotized, and those who have strong fantasy lives and can become lost in an imaginative world, such as a book or movie.

Hypnosis experiments in the 1950s showed the mind to be so suggestible that people would follow instructions against their better judgment.

"6...5... and down we go deeper and deeper. Your body's getting heavier and heavier, relaxing even more now, with every breath that you take. 4..."

“Action seems to follow feeling, but really action and feeling go together.”

PSYCHOLOGIST WILLIAM JAMES

that don’t reflect real childhood behavior, according to child psychologists. As even Freud acknowledged, hypnosis does not actually retrieve memories from early childhood or help in resolving trauma. It does, however, reveal the immense power of imagination and social influence.

THE SOCIAL BRAIN

Most 20th-century research into the brain and behavior focused on the brain as a solitary biological machine, built according to an ancient genetic code. As fascinating as the findings were, they didn’t reach very far into the everyday social and emotional interactions that define our lives and our cultures. Beginning in the 1990s, researchers broadened their study of the links between brain activity and behavior to include social and emotional reactions, bringing neuroscience into social science. They’ve discovered, among other things, where recognition of faces resides in the brain; that physical pain and the pain of social rejection involve the same brain areas; and that loneliness can be inherited. They’ve also found that the social affects the biological. A child’s upbringing can affect his nervous system. Environment and education can help to retrain neurons and rebuild damaged brains. The brain is plastic, responding to both interior and exterior pressures.

» Do I Know You?

One of our most fundamental social actions is recognizing other people’s faces. Studies of both normal and damaged brains have revealed that object recognition, generally, is a complex, highly compartmentalized function of the brain. When you look at an object—let’s say your dog—visual signals from your eyes speed into your temporal lobes, which then begin to classify it. Is it living or nonliving? What color is it? Large or small, moving or still? What sound does it make? What emotions are connected to it? This web of associations and categories rapidly merges into recognition: That’s my dog, Charlie.

Facial recognition falls into its own special category, separate from recognizing other body parts. People

That’s my dog, Charlie

People with prosopagnosia sometimes cannot identify even those they love by seeing their faces.

who have severe difficulty recognizing faces, even after long acquaintance, or who simply can't distinguish them at all, have prosopagnosia, or face blindness. The condition may be genetic in some people and arises after damage to certain areas of the temporal lobes in others. Neuroscientist Oliver Sacks, who wrote a famous account of object blindness in *The Man Who Mistook His Wife for a Hat,* has prosopagnosia himself. In an essay, he recounted how he was unable to recognize his longtime therapist. "I am particularly thrown if I see people out of context, even if I have been with them five minutes before," he wrote. "This happened one morning just after an appointment with my psychiatrist. (I had been seeing him twice weekly for several years at this point.) A few minutes after I left his office, I encountered a soberly dressed man who greeted me in the lobby of the building. I was puzzled as to why this stranger seemed to know me, until the doorman greeted him by name—it was, of course, my analyst." Sacks even has difficulty recognizing himself. "On several occasions I have apologized for almost bumping into a

The prefrontal cortex is the brain's chief executive officer.

large bearded man, only to realize that the large bearded man was myself in a mirror."

Prosopagnosia does not extend to recognition of animals—farmers with this problem have no trouble identifying their individual sheep. And typically, it involves strictly visual recognition, but not emotional recognition, even if the emotions don't consciously register. In one study, patients with this deficit were shown pictures of family and celebrity faces, then asked to name them when shown a second time. Naturally, they were unable to identify any of the faces. Yet electrodes measuring skin conductance response showed that their bodies consistently reacted to the familiar faces. Some part of the brain recognized the faces without the mind's awareness.

From Boston to Beijing, brains are essentially the same.

As Sacks's example shows, people who have even some degree of face blindness may be accused of rudeness, shyness, or general social ineptitude. If you find that you rarely can match a face to a name, or don't recognize someone you just met at a party, you might be relieved to know that like many people you have a mild amount of prosopagnosia. More severe recognition impairments, though, can lead to outright delusions and major

dysfunction. In this, emotional recognition is just as important as visual recognition. Damage to the areas of the brain that supply the emotional associations we have for others can combine with other deficits to create bizarre behaviors. Among them is Capgras syndrome, named for the French psychiatrist who first described it. Patients suffering from this disorder insist that people close to them are actually impostors. They recognize their faces, but are certain that they are merely cleverly disguised strangers, and no amount of rational argument can persuade them otherwise. The syndrome is seen in schizophrenics and Alzheimer's patients, among others. It has sometimes led to violence, as in the case of a man who killed his own father, convinced that the older man had been abducted and replaced by a robot.

THE WALKING DEAD

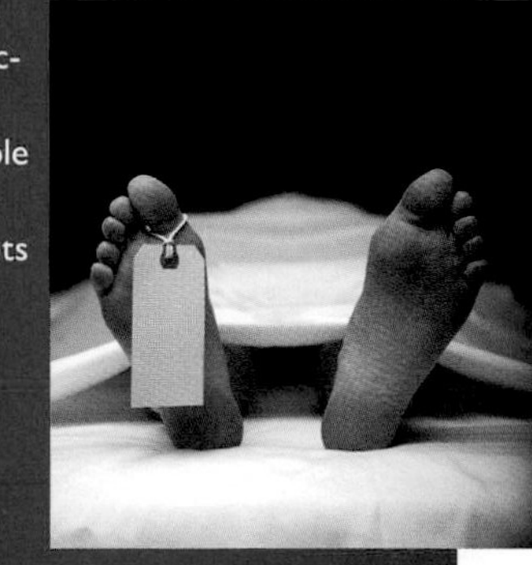

As disruptive as it is not to recognize people who are close to you, it is even more destructive not to know yourself. In a rare counterpart to Capgras and Fregoli syndromes, people with Cotard's syndrome believe that they, or parts of their body, are dead. Cotard's patients often neglect to eat or wash; they may insist on being buried or taken to the cemetery. A patient described in 2008, "Mrs. L.," was admitted to a hospital "complaining that she was dead, smelled like rotting flesh, and wanted to be taken to a morgue so that she could be with dead people."

The condition is associated with a wide range of disorders, ranging from bipolar disorder and schizophrenia to psychotic depression. It can also appear after brain injury, as in the case of a Scottish man who, following a motorcycle accident, became convinced that he had died of septicemia in the hospital and was being escorted around hell by the spirit of his mother. (The mother had in fact taken the man to South Africa.)

The cases are so few, and are connected to so many mental disorders, that researchers have not found one single source of the delusion. However, scans of patients indicate that at least some have impaired functioning in the neural circuits that control visual and emotional recognition of faces. In the case of Cotard's syndrome, the loss of recognition seems to extend to the self as well, with dire results.

The opposite effect is seen in Fregoli syndrome. People with this condition, brought on by damage to the brain's right hemisphere, have an overactive sense of recognition. They are convinced that strangers are in fact people familiar to them, concealed by a disguise. The syndrome is often connected to a paranoiac belief that these familiar people are following the patient around, spying on him and repeatedly changing their disguises. (The syndrome gains its name from Italian actor Leopoldo Fregoli, famed for his ability to quickly change his appearance on stage.)

Recognizing faces, attaching emotions to that recognition, and making judgments about the other person's emotions and personality form a fundamental level of human social interaction. The information processing that goes on in the brain during these moments is a key part of

"Act how you want to feel and you will feel the way you act."

WRITER GRETCHEN RUBIN

cognitive psychology. Social psychologists study the biases, attitudes, and cultural contexts that affect such interactions. In both fields, researchers are increasingly turning to brain imaging to track down the neural pathways that become active when one person deals with another.

» Watching the Brain in Action

To monitor brain activity, the old standby of mind-reading machines, the electroencephalograph, can still be used. The brain is an electrochemical organ, with waves of electrical communication sweeping across it every moment. Using electrodes attached to the scalp, doctors can pinpoint changes in normal wave patterns and even trace those changes to small areas of the brain.

The electroencephalogram (EEG) recording that shows those brainwaves doesn't produce an actual image of the brain.

The more electrodes, the fuller the picture from inside the brain

However, computed tomography (CT) scans, MRIs, and PET scans will do that, and PET and functional MRI images can even track mental activity as it happens. PET scans track the brain's consumption of radioactive glucose that has been injected into the patient being scanned. (The radioactive dose is low and harmless.) More active parts of the brain suck up more glucose, so the machine tracks activity almost as it happens and can map out more and less active regions.

Functional MRIs (fMRIs) produce even more detailed images of brain metabolism than PET scans. MRI stands for magnetic resonance imaging, a technique that registers the signals released by brain cells in a magnetic field. Functional MRIs can identify regions with greater blood flow, hence greater activity, and can narrow those areas down to about 1 millimeter. Research using fMRIs has identified areas of the brain related to very specific functions, such as the facial recognition mentioned above, or the mental mapping shown by the vegetative patient in chapter 1. These scans are measuring proxies, tracking blood flow rather than actual neuron-to-neuron communication. Nevertheless, they have opened up important new fields of study that can link thoughts, emotions, and behaviors to the brain.

» Seeing Ourselves

Using these neuroimaging machines, scientists have cast a light on what happens within the brain as we reflect on ourselves and navigate

( ) ASK YOURSELF

WHERE DO I STAND?

Perhaps you haven't said it out loud, but you probably have a pretty good idea of how your personality and strengths compare to those of other people. Try a simple activity. Rate yourself according to each of the traits below. What percentage do you fall into compared to the overall population? In other words, if you believe you are about as friendly as the average person, you'd rate yourself at 50 percent.

- **Humor**
- **Emotional intelligence**
- **Emotional stability**
- **Driving ability**
- **Friendliness**
- **Honesty**
- **Trustworthiness**
- **Empathy**
- **Logical reasoning**
- **Detecting liars**

How often do you rate yourself above 50 percent? If you usually do, don't worry: In that way, if in no other, you're just like everyone else.

the social environment. Recognition of faces and emotions is basic to our sense of where we fit in to a social world and how we relate to others. So is our conception of our unique self—our assessment of our strengths and weaknesses, our self-control or lack of it, our association with a particular group.

Take something as basic as self-perception. How do we see ourselves relative to others? Research shows that most of us think we're pretty darn good. When asked how accurately positive and negative personality adjectives describe the self, normal subjects judge positive traits to be overwhelmingly more characteristic of their selves than negative attributes. In Lake Wobegon syndrome (from the fictional town where all the children are above average), people consistently rate themselves higher than their peers. Moreover, people typically believe that they are responsible for their own positive characteristics, but that outside forces lie behind their negative ones. As researcher Mark Alicke puts it: "I make myself good; fate makes me bad." Although having an accurate self-evaluation can be useful in making your way through life, maintaining self-esteem and a positive outlook seem to outweigh accuracy for most people.

Neuroimaging studies have been able to link self-evaluation to specific areas of the brain, namely that busy zone known as the prefrontal cortex (PFC). Located in the brain's frontal lobe, just behind the forehead, this region is responsible for many of the processes that we associate with our most evolved selves—our judgment, decision-making, complex thoughts, problem-solving, conscience, and empathy. The prefrontal cortex is the brain's chief executive officer, pulling in information from all the senses and from internal sources as well, including memory and emotional attachment. It regulates behavior and when damaged can lead to a distressing deterioration in self-control and social skills.

We want to believe others always see the best in us.

When people contemplate themselves, the prefrontal cortex becomes active. In fact, self-reflection activates two specific regions, one inside and one just outside the PFC. The medial prefrontal cortex (MPC) processes the actual self-reflective material, the thoughts about the self. The ventral anterior cingulate cortex (ACC), just behind and connected to the PFC, handles the reflection's emotional content. Thus, the thought "I am a strong person" will be understood by the MPC, while the corresponding emotion, perhaps pride, will be supplied by the ACC.

Some studies have found a connection between this emotional self-recognition and depression. People with

We know which part of the brain exerts self-control—but can't always activate it when needed.

depression may have reduced activity in the ACC, indicating that although they may understand the content of their self-reflective thought ("I am a strong person"), they can't appreciate its emotional meaning. When electrical impulses are sent into the ACC via deep brain stimulation, the impulses seem to decrease symptoms of depression in patients who don't respond to standard psychological therapy.

»Self-regulation

Late at night, in the throes of a post-holiday diet, you open up the refrigerator and see that last piece of apple pie. Instantly, you want to pull it out and eat it. Hunger pangs prod you and you can mentally taste the sweet deliciousness of the dessert—but you close the refrigerator door and move on.

How did you resist such a strong, almost primal, urge? You possess self-regulation, a skill that allows you to delay gratification, control appetites and impulses, and persevere toward attaining goals (see chapter 6). It's a tough and important personal quality that allows you to succeed in life. Its failures are equally important. Domestic violence, drug abuse, binge eating, and other disorders are persistent societal problems stemming from failed self-control.

Neuroimaging studies link activity in the prefrontal cortex, specifically the anterior cingulate cortex, to the wide array of mental processes that work together to give us self-control. These mental activities include decision-making; choosing appropriate behavioral responses from among many alternatives; monitoring of performance; processing conflicts; identifying errors; assessing reward and punishment; perceiving social pain; and more. Dysfunction in the ACC is connected to many mental disorders. Among them are obsessive-compulsive disorder (OCD); autism; schizophrenia; and Tourette's syndrome, a condition in which people are unable to control their own tics, noises, or words.

Self-control applies not only to behavior, but to thoughts as well. One of the most painful aspects of OCD, for instance, is the inability to control one's own thoughts. Post-traumatic stress disorder, attention deficit/hyperactivity disorder (ADHD), and depression are also marked by unwanted, intrusive thoughts that are difficult to dismiss. The same areas of the brain are activated when we suppress thoughts as when we suppress behaviors. A study of college students, for instance, asked them to pick a particular thought that was meaningful to them (such as "a phone call with a distant girlfriend") and then either suppress that thought, clear the mind of all thoughts, or let the mind wander freely. The ACC lit up in the fMRI scans when the students attempted to suppress a single idea. Scattered regions of the brain, including the ACC, became active when they tried to clear their minds completely.

Emotional control also seems to reside in the prefrontal cortex.

"I don't sing because I'm happy, I'm happy because I sing."

PSYCHOLOGIST WILLIAM JAMES

When we think about our own emotional states or those of others, when we watch sad movies or look at frightening photos, the ACC and related areas are involved. So is the amygdala, which processes emotions such as fear. Researchers have found that people can consciously change these emotional reactions by reappraising them. Shown a frightening photo, for instance, after an initial shocked reaction a person can think again, deliberately distancing herself from the emotions involved. As she does this, the ACC control center becomes more active and the amygdala is deactivated, defusing the fear.

The inability to regulate emotions is part of many disorders, including depression and anxiety, as well as aggressive, violent behavior. Doctors hope that brain studies and techniques such as reappraisal will lead to better therapies for these serious problems. The rather amazing fact that we can alter our brain chemistry simply by thinking about it has big implications for psychology in general.

» In-Groups and Out-Groups

As we've seen, recognition of other people is deeply rooted in the brain. Particularly in early human history, the ability to distinguish friend from foe, neighbor from stranger, was a vital survival skill. The brain devotes considerable resources to understanding the relationship of others to the self. Who are these people around me?

FRIEND?

OR FOE?

The human mind often categorizes the social world into "us" and "them."

How do they relate or not relate to me? Should I be worried about their presence in my life? Should I be happy they are here? Meeting another person, we're motivated to answer two questions in short order. The first is "Friend or enemy?" The second is "Are their goals similar or dissimilar to my own? Do they want what I want?"

Like it or not, this type of processing leads to a rapid, clear mental distinction between "us" and "them." Quick judgments about in-groups and out-groups, and the irrational prejudices that can accompany them, may be an inevitable part of human nature. Categorizing people into social groups allows us to simplify the social world and generalize our existing knowledge about certain groups and new people. We instinctively identify the group that we are a part of (i.e., us) and those that we are not part of (i.e., them). Research has found that this group membership actually alters the way the brain is wired. Thus, the brain tends to process information in a way that favors members of our in-group (i.e., us) and punishes members of our out-group (i.e., them). If you find yourself making snap judgments about people based on quick impressions, you can realize that the behavior has ancient roots—but you can also consciously be aware of it and overcome it.

When we think about prejudice and discrimination, the examples that jump to mind are the obvious ones that use visual cues: racial prejudice based on skin color, or religious discrimination tied to specialized clothing. But the preference for the in-group versus the out-group goes much deeper than that. Moreover, it can be formed almost instantaneously. The simple fact of being separated into groups can trigger the reaction.

Researchers tested this in an experiment involving British schoolboys. The students were brought together and given a meaningless task: estimating the number of dots on a screen. Then they were divided into arbitrary groups based on their supposed performance. Boys in each group were asked to award points, worth a small amount of money, to others who were identified as being either in their own group or in the other group. Overwhelmingly, the boys favored their own groups, random as they were, giving them much more money.

The teenage experience of in-groups and outcasts actually shows up in the neural activity of the brain.

Is this tendency hardwired into our brains? Some studies indicate it is. The basic in-group, out-group distinction can be applied to almost any set of people, depending on social surroundings. It isn't fundamentally connected to race or appearance. One study, for instance, randomly assigned white participants to a mixed-race in-group, had them briefly learn the members of their new in-group, and then presented them with in-group and out-group faces during an fMRI scan. Participants had greater amygdala activity in response to members of their novel, mixed-race in-group than they did to out-group faces—in other words, they had stronger emotional reactions to their group than to the other. This in-group bias in neural processing occurred within minutes of group assignment.

FOCUS

THE ANGRY STRANGER

Study after study has shown that we recognize people in our own ethnic groups more easily than those in other ethnic groups—the unfortunate "they all look the same to me" syndrome. Psychologists Joshua Ackerman and colleagues in fact used this phrase in the title of their article "They All Look the Same to Me (Unless They're Angry)." In their study, they presented white students with quick glances at both white and African-American faces. Then they asked them to identify previously viewed faces. When the faces had a neutral expression, the results were predictable: The students recognized white faces more accurately than African-American ones. But when the faces were angry, the white students recognized African-American faces at the same rate as white ones.

From an evolutionary perspective, this makes sense. Early humans benefited from distinguishing their own groups from others, and from reading social and status cues in the expressions of their own group members. But different motivations come into play when those expressions are angry. The threat level from angry strangers is considerable, which means that our survival is bolstered by remembering an angry, and unfamiliar, face.

ROOT, ROOT, ROOT FOR THE HOME TEAM

Few domains have more obvious, vocal, and enthusiastic examples of in-group/out-group aggression than sports. Sports fans actively deride the opposing team and its fans and cheer when they suffer misfortune. This lack of empathy—indeed, outright aggression—shows up in the brain. Hooked up to fMRI machines, avid Boston Red Sox and New York Yankees fans were studied as they watched their hated rivals either win or lose. Areas of the brain linked to pleasure lit up when the rivals lost, even if they lost to another team (say, the Baltimore Orioles).

Nor is the sporty schadenfreude limited to the team themselves. Soccer fans show increased activity in the brain's reward center when they see a rival team's fan get a painful electric shock. The greater the activity in these pleasure centers, the more likely the fan was to report a desire to hurt the other teams' fans, and the less willing he was to relieve the other fan's pain by taking on some of it himself.

LOVE AND OTHER EMOTIONS

Instant antipathy is one thing, but there is remarkably little consensus among psychologists about the nature and course of long-term romantic love. Many believe that love fades over time. Others think that it evolves from passionate attraction into friendship or companionship. Freud was convinced that any long-lasting passion was pathological. However, advocates for eternal romance have found some support in neurological studies. In one, women who reported still being in love with their longtime husbands were shown photos of their spouse and of three other acquaintances of varying familiarity. Only when viewing their husband did a variety of brain regions light up; among them, dopamine-rich reward areas associated with pleasure, "liking," and early romantic love; our old friend the ACC; and regions linked to maternal pair bonding. Compared to those newly in love, the brains of long-term partners had a much wider range of response to the loved one. Other research suggests that the presence of a loved one is not just pleasant, but calming, as the brain regions involved help to modify anxiety and pain.

Many people have had the unpleasant experience of seeing a loved one in pain. Neural studies show that the sensation you have of actually experiencing the other person's pain is a real one. In one study, researchers applied electrodes to a participant's ankle and showed

Neuroscientists still don't agree with one another on what love is.

him or her a series of images while the subject held hands with either a complete stranger or a close friend. The other person also had electrodes around his ankles. Certain images indicated the possibility that the participant might receive a slight shock, while others indicated that the other person (friend or stranger) might receive the same slight shock instead. Using functional magnetic resonance imaging (fMRI), researchers compared brain activations for self-focused threat (e.g., "I am about to be shocked") to those for threats to a familiar friend or an unfamiliar stranger. The results: Areas of the brain that register threats to the self also reacted when a friend was threatened, but not when a stranger was involved. One of the defining features of human social bonding may be increasing levels of overlap between neural representations of self and other. Quite simply, the closer I am to you emotionally, the more my brain reacts to a threat to you as if it is a threat to me.

» Emotions on the Left and Right

Our emotions play a central role in how we respond to the social world, but just what emotions are, and where they begin, has long been debated. William James, one of the founders of modern psychology, believed that emotions start with the body and travel to the brain. "We feel sorry because we cry," he wrote, "angry because we strike, afraid because we tremble."

Others disagreed. Physiologist Walter Bradford Cannon,

Brain or body? Studying the mind often presents a chicken-or-egg situation.

working in the 1920s, thought James's theory was too dismissive of the brain. In his view, both brain and body react simultaneously to a stimulus. The angry dog barks at us, and the signal travels both to the sympathetic nervous system, raising our heartbeat, and at the same time to the brain's cortex, registering the threat and making us aware of fear.

By the 1960s, cognitive science had entered the fray. Scientists who studied information processing in the brain proposed the "two-factor theory" of emotion. According to this scenario, when we encounter the angry dog, our body reacts (the heart pounds) and our brain interprets this physical reaction, recognizing that it signals fear. Only then do we consciously register the emotion of fear.

Modern researchers, armed with brain-scanning technology, are attempting to resolve this chicken-or-egg (or dog-or-brain) dilemma by observing just what happens, and where, in the brain when it experiences emotions. The evidence so far seems to support both the James "body-first" theory and the two-factor theory. People exposed to rapid, subliminal images will react emotionally to those images without conscious awareness of their content—body first. However, people who have been physically stirred up by an injection of epinephrine, but told that the drug would arouse their emotions, feel much less emotional than those who were not told of the drug's effects. Their brains have interpreted the drug-based arousal as unimportant and their emotions have adjusted accordingly.

It appears that some sensations are passed directly from the ear or eye to the thalamus and the amygdala, which might produce an instant emotional reaction. However, the amygdala typically sends on its response to the prefrontal cortex, our executive center, which appraises the situation and regulates emotions. Different parts of the PFC are linked to different kinds of emotions, with a distinct left-brain/right-brain split. In experiments in the 1960s, investigators injected Amytal, a

"I make myself good; fate makes me bad."

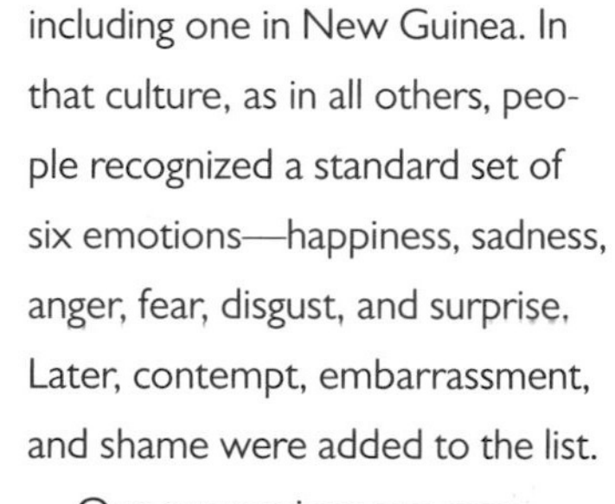

RESEARCHER MARK ALICKE

barbiturate, into one of a subject's internal carotid arteries, which temporarily suppressed the activity of one brain hemisphere. When the left hemisphere was suppressed and just the right was active, the subject felt depressed. When the right hemisphere was suppressed and just the left active, he felt euphoric.

The left-right difference has been supported in other studies as well. People with damage to the left prefrontal cortex are more likely to feel depressed. In scans, people who show more activity on the left side of the PFC are more likely than others to be happy, curious, and hopeful. Those with greater right-side activity are more apt to experience anxiety and mood disorders. The left side responds strongly to pleasure. Even at two days old, infants show more action in the left PFC when a little sugar is placed on their tongues. The same areas light up in adults when they view luscious desserts.

Emotions are expressed on the face and in behavior, and the human mind is exquisitely sensitive to these indicators. Facial expressions, in particular, are universal communicators, conveying the same emotions in every culture. In a famous study, researchers showed photographs of faces to people in different cultures,

including one in New Guinea. In that culture, as in all others, people recognized a standard set of six emotions—happiness, sadness, anger, fear, disgust, and surprise. Later, contempt, embarrassment, and shame were added to the list.

Our expressions are spontaneous, but can be modified by social situations. A study of Olympic judo athletes compared their instant expressions at the end of a match with their expressions on the medal podium. Just after the match, 13 out of 14 gold medalists had what psychologists call a "Duchenne smile"—that is, a genuine smile (named after physician Guillaume Duchenne) that pulls up the corners of the mouth and raises the cheeks. None of the second-place finishers smiled, though some showed sadness or contempt. On the podium, the gold medalists all had genuine smiles, while silver medalists mustered a few—some genuine, some forced, and some a blend of smiling and sadness.

Bronze medalists, on the other hand, seemed much happier than silver medalists, with more genuine smiles. People are apparently happier when they know they've just made it onto the medal stand than when they consider that they've just lost out on the gold.

Not only do emotions cause expressions, but expressions cause emotions as well. Supporting William James's assertion that "we feel sorry because we cry," researchers have found that people who stretch their faces into shapes that suggest emotion—brows pulled together in a frown, or mouths stretched widely in a smile shape—actually experience those emotions to some extent. Just holding a pen in the teeth to stretch the mouth into a smile shape makes cartoons seem funnier. Contracting the eyebrows makes sad movie scenes seem sadder. The effect is low-key and doesn't outweigh other sources of emotion, but it does demonstrate the complex feedback system that connects body and mind.

CHANGING YOUR MIND

From Boston to Beijing, brains are essentially the same, with the same regions devoted to speech, memory, motion, vision, and an array of other functions. It used to be thought that little could affect the predestined spread of the brain's billions of neurons. Furthermore, damage to the brain was also seen as irreversible. Unlike some other cells, damaged or cut nerve cells usually do not regrow. But today it is clear that the brain is considerably more malleable—more

At times, the face expresses more than words could ever utter.

plastic—than once believed. Not only can it respond to and, to some extent, recover from physical damage, but also it can be shaped simply by experience and intention. And what shapes the brain shapes the mind.

In many cases, the brain can compensate for damage in one area with new learning in another. Stroke victims often recover some lost abilities as other areas of the brain pick up the slack for impaired sections. Control of the fingers will expand into the visual area in blind people learning Braille. This sort of plasticity is especially apparent in a child's developing brain. Children who have undergone hemispherectomies (removal of one brain hemisphere) to prevent seizures will retain or regain many of their skills and personality traits as the remaining hemisphere takes over.

In 2014, Chinese doctors reported the remarkable case of a 24-year-old woman who checked into the hospital complaining of balance problems. This was hardly surprising, since scans revealed that the woman had been born without a cerebellum, the important, neuron-dense region of the brain that coordinates movement. She reported that she had learned to walk at age seven and had slurred speech until about six (the cerebellum plays some part in speech, as well). It had taken fewer than eight years for her remaining brain to somehow pick up the complex functions of the cerebellum, allowing the woman to lead a normal life, get married, and have children.

Dramatic cases of brain compensation like this are relatively rare, but the ways in which experience affects the brain and mind are myriad. Professional musicians have larger gray matter areas devoted to motor control, auditory processing, and visual-spatial information than nonmusicians. A study of Westerners who practiced meditation at least once a day for 40 minutes found that, compared with nonmeditators, they had thicker cortical regions devoted to sensory, cognitive, and emotional processing.

Although people are born with certain personality traits and approaches toward life, these too can be consciously modified with meditation. People who took an eight-week mindfulness training course showed more activity

RELAXING THE FACE AND MIND

Many meditation techniques involve focusing on and relaxing specific muscle groups, until the whole body is at ease. Research showing that the brain reacts to stress in the body supports the idea that a more tranquil body leads to a more tranquil mind.

This short exercise will give you a taste of these techniques. Pause for a few seconds between each step, and stop if you experience any discomfort.

» Find a peaceful location. Sit or lie down.

» Take a deep breath and exhale.

» Raise your eyebrows high, hold them for ten seconds, then release.

» Squeeze your eyes tightly shut, hold for five seconds, then release.

» Smile broadly, hold for five seconds, and release.

» Slowly tilt your head back to look at the ceiling. Hold for five seconds and release.

in the left sides of their brains, resulting in less overall anxiety. Those who practice compassionate meditation (see chapter 8, page 251) have more active mental circuits in regions related to empathy. If you're not happy with the way your brain is working, there are ways you can redirect it and actually alter its shape.

»Therapy and the Brain

We can change our brains through repetitive learning and through meditation. We can also quite literally change them with therapy. Practitioners of cognitive behavioral therapy (CBT), a widely practiced, targeted approach to dealing with emotional and behavioral problems, have been eager to know whether their treatments have real results in the brain. So far, the answer seems to be yes.

THOUGHTS
BEHAVIOR
EMOTION

In cognitive behavioral therapy, a person reviews thoughts and actions as well as feelings.

The brain's 90 billion neurons create complex webs of connections.

Stress, for instance, registers in the brain's cortex. In one study, people who underwent two days of cognitive behavioral therapy—cognitive restructuring (changing unhelpful thought patterns to more helpful ones), problem-solving, self-instruction, and progressive muscle relaxation—showed a significantly calmer response in their brains to a stress test afterward. The therapy's effect was still visible four months later.

Exposure treatment for phobias, another kind of CBT, also leads to alterations in the brain. The treatment involves gradually exposing the phobic person to the object of their phobia, bringing it closer and making it more real over time. The phobic person learns that no harm comes to them in the presence of the feared object, and their stressed, fight-or-flight reactions diminish. Spider phobias, among the most common, have been successfully treated this way, and researchers followed one group of subjects to see what happened within their brains in the process.

Before their therapy, the spider-phobics' brains showed a range of activity when they looked at spiders, including in the prefrontal cortex. This suggested to researchers that they had to call on self-regulation to tamp down their fears. After CBT, abnormal activity in this prefrontal area had vanished. The therapy had reduced the participants' fears so much that the brain no longer needed to step in. Four weeks of mental exercises had rewired the brain's connections.

Many questions remain unanswered about the brain and its connection to the mind and behavior. Those who study these questions are optimistic that the answers they find will not only add to our knowledge bank, but will also allow psychologists to directly link therapy to the brain, freeing people from their most distressing thoughts and behaviors.

» CHAPTER THREE «

HOW WE GROW

In 1992, University of Arizona psychologist Karen Wynn conducted an ingenious experiment with a very young group of subjects: five-month-old babies. In a kind of arithmetic theater, she seated the infants in front of a display area that held a single object. After a screen rose up to cover the object, an experimenter appeared with a second object and clearly placed it behind the screen. Then the screen was pulled away to reveal either the predictable result—two objects—or a surprising one—one object. The infants stared longer at the surprising scene. They showed the same reaction to impossible subtraction scenes as well, looking longer at a result when the experimenter showed them two objects to start, removed one behind the screen, and then took away the screen to reveal two objects remaining. Knowing that babies will look longer at unexpected events than at expected ones, the experimenters had to conclude that the small spectators were doing some basic math in their heads. One plus one should equal two; two minus one should equal one; and the babies were surprised when they didn't.

"This indicates," wrote Wynn, "that infants possess true numerical concepts, and suggests that humans are innately endowed with arithmetical abilities." And so one more shot was fired in the nature vs. nurture battle that has defined psychology since its earliest days.

THE BLANK SLATE

We can see the biological basis of our behavior not only through brain scans or evolutionary studies, but also in the growth of every child. People follow predictable stages of psychological development not just during childhood, but also, as

Many psychologists and philosophers used to believe that people were born as a "blank slate" without any intrinsic traits or abilities.

psychologists are now discovering, even during adulthood. This wasn't always acknowledged. Through much of the 20th century, proponents of the blank slate theory of human development held the field. Behaviorist psychologists, in particular, believed that the environment shaped virtually all of human development. Conditioned reflexes and learned responses determined our behavior, our skills, and our fate.

Among the first to espouse this view was psychologist John B. Watson. Watson, a pioneer in behaviorist studies, was a firm blank-slater. "Give me a dozen healthy infants, well-formed, and my own specified world to bring them up," he wrote, "and I'll guarantee to take any one at random and train him to become any type of specialist I might select—doctor, lawyer, merchant-chief, and yes, even beggarman and thief, regardless of his talents, penchants, tendencies, abilities, vocations, and race of his ancestors."

While a professor at Johns Hopkins in 1919 and 1920,

"Education is what survives when what has been learned has been forgotten."

PSYCHOLOGIST B. F. SKINNER

Watson with his colleague Rosalie Rayner performed a famous—indeed infamous—experiment on a single eleven-month-old infant, identified in his research as Albert (and now more commonly known as Little Albert). Watson believed that many emotional responses were the result of conditioning. "In infancy," he wrote, "the original emotional reaction patterns are few, consisting so far as observed of fear, rage, and love." To show that a particular fear could be learned, the experimenters showed placid Albert a white rat, a rabbit, a dog, masks with hair, and other furry objects. Then they began to teach Albert to fear the rat. Every time they presented it to him, an experimenter would bang loudly on an iron bar just behind the infant's head, shocking and frightening him. Soon enough, just the sight of the rat was enough to scare little Albert. He quickly generalized this fear to other animals, such as rabbits and dogs, and even animal masks, but not to objects such as cotton wool.

Such an inhumane experiment could not be conducted today. Watson intended no cruelty by it, however. Primarily, he wanted to refute Freudian theories that childhood sexual conflicts were at the root of phobias. Emotional disturbances, he wrote, must instead be traced to "conditioned and transferred responses set up in infancy and early youth."

Experiments with a white rat showed how a conditioned response develops.

» Conditioning and the Learned Life

Little Albert learned to fear fluffy animals through conditioning, a principle that has been part of psychology since the days of Ivan Pavlov and his salivating dogs. Watson showed its efficacy with one unfortunate child. B. F. Skinner made it, and the behaviorist approach, famous.

Skinner, who was a major figure in experimental psychology from the 1930s to the 1970s, is best known for his theories of operant conditioning. In classical conditioning, an animal or human makes an association between one stimulus (say, a bell ringing) and another (food). An animal's original, unconditioned response is the one that comes naturally and spontaneously—say, drooling while it eats. But if a bell is

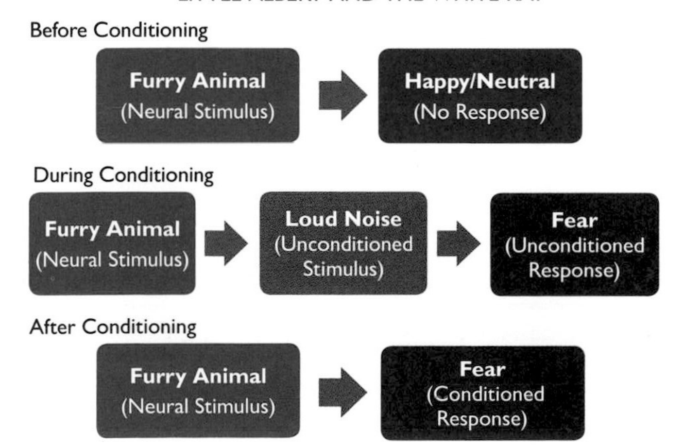

FOCUS

B. F. SKINNER

Burrhus Frederic Skinner, born in the small town of Susquehanna, Pennsylvania, in 1904, originally dreamed of becoming a novelist, but concluded that he didn't know enough about human behavior. He turned to psychology to give him the insights he lacked.

By the time he received his doctoral degree from Harvard, Skinner had already built the animal-training operant chamber that would make him famous. After stints at other universities, he joined the faculty at Harvard and went on to reign as the era's most prominent psychologist.

In his 40s, Skinner combined his first ambition with his later career and wrote a novel about a utopian community run on behaviorist principles. *Walden Two's* anti-authoritarian society was the inspiration for a number of real-life experiments in utopias.

Skinner died of leukemia in 1990, leaving behind a field in dispute and some undoubtedly major contributions to our understanding of how animals and human beings learn and grow.

rung each time food is about to be served, then in time, the animal develops a conditioned response to the stimulus by drooling when the bell rings.

Operant conditioning takes this one step further into active, learned behavior. In operant conditioning, an organism learns to associate its behavior with reward or punishment. Over time, it increases the rewarded behavior and decreases the punished actions; in other words, the rewarded behavior is reinforced. The animal has learned to operate on its environment to achieve a goal. The now-iconic scene of a rat in a box, pressing a lever for food, is an example of operant conditioning. It was Skinner who invented that box, now known as an operant chamber or Skinner box.

Skinner worked not only with rats, but also with other animals, most famously pigeons. Using operant conditioning, he taught a pigeon to play the piano by pecking the keys. In World War II, he even developed a pigeon-guided missile system (sadly, never used). In his view, animals were all the same and a human was just another pigeon, a creature whose behaviors were

Babies as young as 42 minutes old can imitate facial expressions.

reinforced in the same ways. Attributing behavior to inner thoughts and human nature was "prescientific," according to Skinner. "Thinking is behaving," he said. "The mistake is in allocating the behavior to the mind."

Skinner's behaviorist approach was hugely influential. Today, the basics of operant conditioning are applied to practical problems in a variety of settings, from schools to prisons. Conditioning therapies to treat phobias have been highly successful (see chapter 2, page 70). But, as with most such bold theories, time and further research have tempered or contradicted many of Skinner's assertions. For one thing, we've amassed considerable evidence for innate behaviors and motivations. Rats in mazes may respond to rewards, but they will navigate even without them and seem to form mental maps on their own.

Psychologist B. F. Skinner built a glass-walled air crib for his infant daughter.

Animals, including humans, are also biologically predisposed to certain behaviors and not others. You can train a pigeon to peck in order to get food, because pecking for food is a normal pigeon behavior. However, no amount of rewards will reliably train it to flap its wings for food. This just doesn't come naturally to a pigeon.

Without denying that people learn and develop in response to their environment, psychologists studying infants and children began to turn away from strict behaviorism and back toward the idea of innate abilities and stages of development. Two thinkers were crucial in this effort: Jean Piaget and Noam Chomsky.

STAGES OF DEVELOPMENT

Swiss psychologist Jean Piaget was a contemporary of Skinner's, but his understanding of childhood development could hardly have been more different. In Piaget's studies of infants and children, inspired by his own family, he concluded that all children progress through innate stages of development. Each stage represents a new level in understanding their environment. "Children are active thinkers, constantly trying to construct more advanced understandings of the world," he wrote. According to Piaget, two processes drive a child's progress through learning. The first is assimilation: interpreting new experiences in the light of the world we understand.

The second is accommodation: changing and expanding our worldview to incorporate the new information.

Piaget believed that all children grow through four stages of development as they interact with the world. They are:

• **Sensorimotor,** from birth to almost two years. In infancy, a child has no sense of past or future, but relates to the world in purely physical terms. He learns by touching, looking, tasting, and hearing. During this stage babies develop stranger anxiety, becoming distressed if handed to an unknown person. They also acquire object permanence. To most young infants, a rattle hidden beneath a cloth ceases to exist. Past about six or eight months, this changes. A child will notice and remember the location of the hidden object, lifting the cloth to reveal the toy.

• **Preoperational,** from two to about six or seven. Children in this stage are learning to represent the world in their heads, through language and images. In this phase, they are naturally egocentric, possessing little understanding of other people's minds. They are literal, but not rational. For instance, the preoperational child doesn't grasp the conservation of volume. She'll believe that a tall skinny glass holds more milk than a wide, short glass.

• **Concrete operational,** ages six or seven to twelve. In these years, according to Piaget, children begin to understand mathematical concepts and the use of symbols. Truly abstract reasoning is still difficult, but they can now comprehend the conservation of materials and how to classify and organize things.

• **Formal operational,** twelve onward. Abstract thinking and an understanding of hypothetical situations mark this stage as children move into adolescence. They can work out puzzles

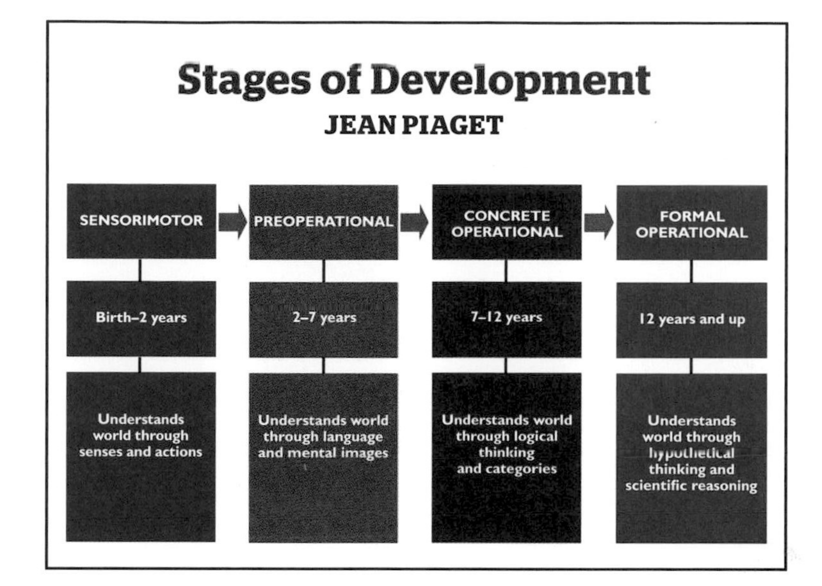

"It is with children that we have the best chance of studying, the development of ... knowledge."

PSYCHOLOGIST JEAN PIAGET

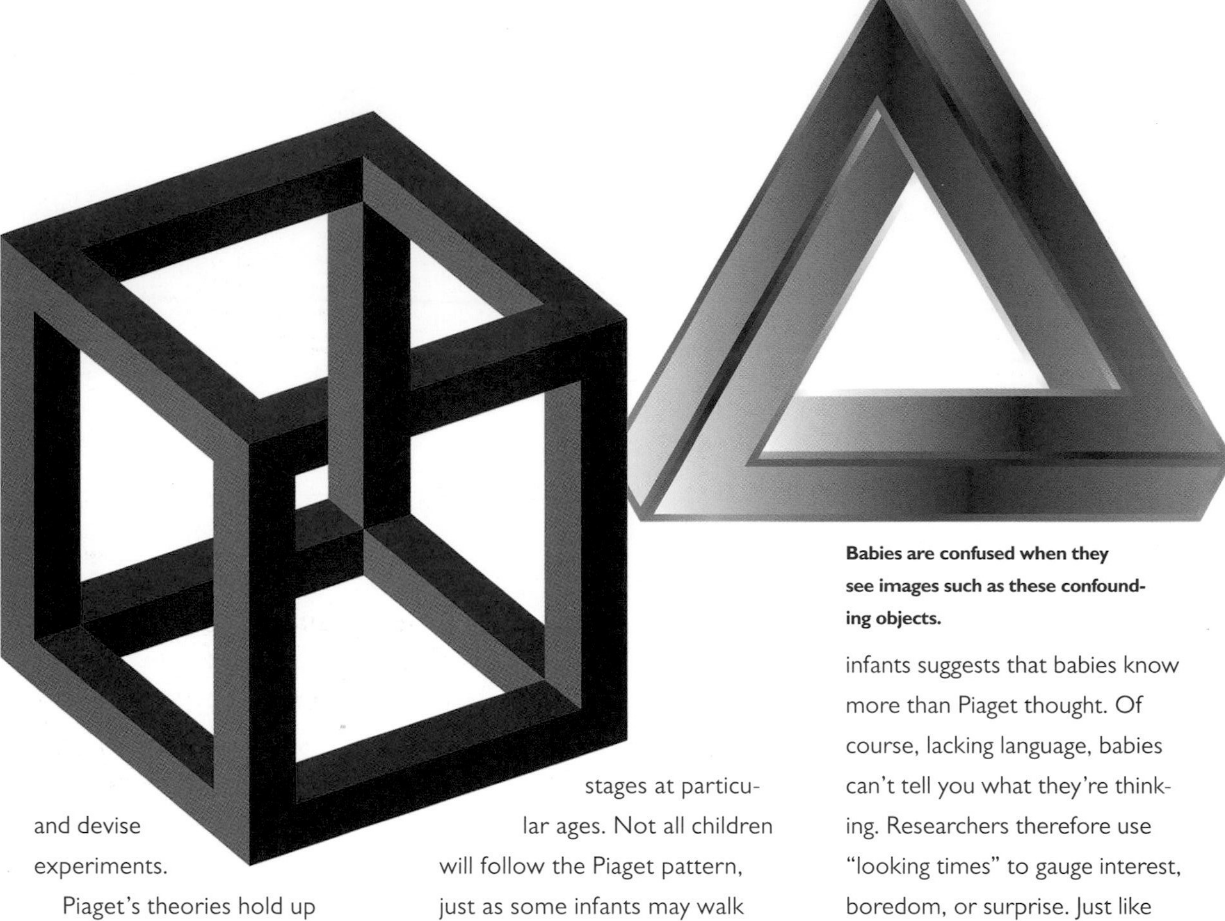

Babies are confused when they see images such as these confounding objects.

and devise experiments.

Piaget's theories hold up pretty well. The stages of development he described have been verified in countless studies of infants and children. However, contemporary psychologists now believe that development typically proceeds fluidly, without strict demarcations between stages at particular ages. Not all children will follow the Piaget pattern, just as some infants may walk without learning to crawl. Logical thinking, too, may be less important than Piaget thought, trumped by the development of social awareness.

»What Do Babies Know?

New research into the minds of infants suggests that babies know more than Piaget thought. Of course, lacking language, babies can't tell you what they're thinking. Researchers therefore use "looking times" to gauge interest, boredom, or surprise. Just like adults, infants will grow bored with an unsurprising scene and look away from it. Unexpected sights attract their attention, and they stare longer at them. This simple technique is what led scientists to believe infants have a feeling for math, as described at

"You're never too old, too wacky, too wild, to pick up a book and read to a child."

WRITER AND CARTOONIST DR. SEUSS

the beginning of the chapter. It has also been used to challenge Piaget's idea that understanding object permanence begins around six months. Some very young babies will briefly look for a vanished object in the place where they saw it hidden.

Four-month-old infants also stare longer at "impossible" objects, such as Escher-style cubes, suggesting that, like adults, they are surprised by the pictures and are trying to understand them. Babies also have an early, sophisticated emotional awareness. Within nine months, they can associate happy facial expressions with a happy tone of voice (showing surprise when an expression and voice don't match).

In fact, infants are born with certain abilities. They are born able to suck, grasp, and focus on objects eight to twelve inches away—the typical distance from a nursing mother's face. They can tell the difference between human and nonhuman faces and voices. They turn toward their mother's face, voice, and smell from birth. Babies as young as 42 minutes old will imitate facial expressions; one-month-old infants will reliably stick out their tongues when they see someone else do it.

It's increasingly clear that social awareness is a key part of child development. Some of the most intriguing studies in this area have to do with the "theory of mind." In the midst of Piaget's preoperational stage, during the preschool years, children begin to understand that other people have their own minds and see things in different ways. This insight typically occurs around three and a half to four and a half years old. Before reaching this stage, a child believes that what she knows, everyone knows. Show her a box with crayons on the label and then reveal that it contains pencils, and she'll be surprised. Ask her what another child, who had never seen the contents, would think was in the box, and she'll answer "pencils." She cannot understand that others might hold false beliefs.

Children with autism often have trouble with these sorts of experiments. Mind-blindness, or the inability to imagine another person's mental state, is characteristic of autism spectrum disorders. People with autism struggle to understand facial expressions or empathize with others, and this might have a biological basis.

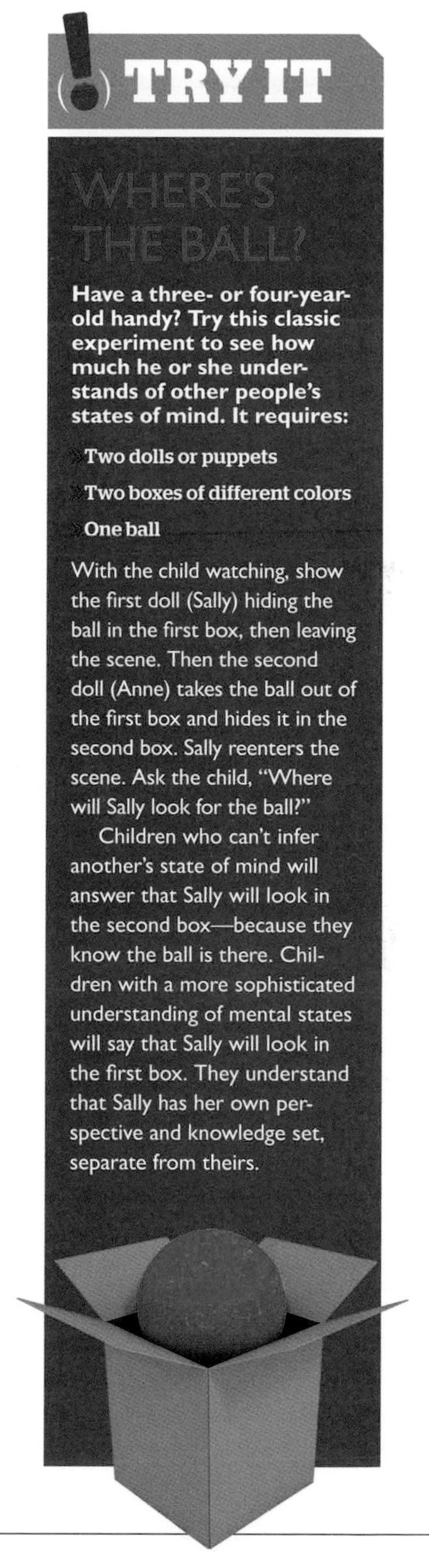

TRY IT

WHERE'S THE BALL?

Have a three- or four-year-old handy? Try this classic experiment to see how much he or she understands of other people's states of mind. It requires:

- **Two dolls or puppets**
- **Two boxes of different colors**
- **One ball**

With the child watching, show the first doll (Sally) hiding the ball in the first box, then leaving the scene. Then the second doll (Anne) takes the ball out of the first box and hides it in the second box. Sally reenters the scene. Ask the child, "Where will Sally look for the ball?"

Children who can't infer another's state of mind will answer that Sally will look in the second box—because they know the ball is there. Children with a more sophisticated understanding of mental states will say that Sally will look in the first box. They understand that Sally has her own perspective and knowledge set, separate from theirs.

Studies of brain function in those with autism show impaired communication among some regions of the brain, areas that are normally active when we try to understand another person's viewpoint. These deficits support the idea that the fundamental social skill of empathizing with others is hardwired into the brain and will emerge at around four years old in a typical child. They also suggest that mental abilities develop as separate modules in the brain. An autistic child who can't decipher facial expressions may still have normal, even excellent, skills in math or science.

» Attachment

A child's emotional growth also proceeds in predictable stages, albeit stages that are strongly influenced by parenting styles. We've seen that even newborns are attuned to a mother's voice and scent. Infants are keenly interested in the people around them—their expressions, their emotions, their reactions. Over the course of their first year, they become increasingly attached to their parents, a bond that reaches its peak around 13 months. At that age, a baby will resist going to a stranger's arms, wailing and pulling back from someone they might have welcomed just weeks before.

The need for touch and physical reassurance is built into the infant psyche. University of Wisconsin psychologist Harry Harlow showed this in dramatic fashion with his experiments of the 1950s that placed infant monkeys with artificial monkey "mothers." Baby monkeys raised with two mother figures—one a chilly wire cylinder with a feeding bottle, the other a soft, cloth-covered form without a bottle—strongly preferred the comfy mother. They clung to her even while stretching over to the wire figure to feed. Like human infants, they used the soft mother as a secure base, leaving to explore their surroundings and then returning for reassurance.

Raised only with artificial mothers, the unfortunate young monkeys became more insecure and more easily frightened than their live-mother counterparts. Human children, too, are shaped by parenting styles. Children raised by sensitive adults, caretakers who notice them and respond consistently to their needs, will in general become more secure, successful, and independent adults. However, even infants who have suffered early deprivation can recover well if given the chance early on. Infants from troubled homes who are adopted before 16 months or so bounce back well.

» The Growth of Language

Every parent knows that the lightning-fast growth of language in a toddler is a marvel. The question of whether language

Attachment to parents and fear of separation peak in the early toddler stage.

is learned from scratch in each child is an old one, and it's key to understanding how much biology contributes to development. The mystery has intrigued the scientifically minded for millennia. Greek historian Herodotus tells us that the Egyptian pharaoh Psamtik I, in order to discover humanity's original tongue, gave two infants to a shepherd to be raised without speech. Supposedly, the first word uttered by one baby was "bekos," the Phrygian word for bread. Therefore, concluded Psamtik, Phrygian must be the world's first language.

We can probably rule out Phrygian as an ur-language (or original language), but we know now that speech is indeed a universal and inborn human trait. By four months, babies can match speech to lip movements; by ten months, their own babbling uses sounds matching the parents' native language. From one to three years, language expands at an astonishing rate, growing from one-word exclamations to complex sentences. Through age seven, a child is a language prodigy, soaking up new words and new languages with ease. After that age, language acquisition drops off. Learning a second language becomes harder, as can be seen in immigrant households, where the parents' new language is awkward and accented, but the children are fluent.

Studies of how language develops in individuals and in cultures have landed some punishing blows against the behaviorist idea that all learning comes from the environment. Linguist Noam Chomsky has been one of the heaviest hitters. In a 1959 review of B. F. Skinner's book *Verbal Behavior,* he pointed out that children understand and produce complex sentences that they've never heard, following rules of grammar they have never been taught. "The fact that all normal children acquire essentially comparable grammars of great complexity with remarkable rapidity suggests that human beings are somehow specially designed to do this," he wrote. Deaf children

FOCUS

THE WILD CHILD

On rare occasions, children have been discovered who were apparently raised with very little human contact—raised by wolves, as folklore would have it. These sad cases fascinate linguistic scholars, providing as they do a natural study in when and how language develops in the absence of outside influences. Confirmed cases of feral children are few, but a close modern substitute is the tragic case of Genie, an isolated girl found in Los Angeles in 1970. Social workers rescued her from her abusive parents, who had imprisoned her in a closed room in silence from the time she was a toddler. At 13 years old, she could not speak or even stand erect. With care, she became sociable, healthier, and the subject of much interest among linguists, who studied her for several years to see if she would develop grammatically complex language just as younger children do.

She did not. Although in time she was able to put together phrases such as "Want more soup," her language never grew beyond the level of a typical two-and-a-half-year-old child. Genie's experience suggested that linguists were right when they theorized that there is a critical developmental period for language.

Children who learn gestural speech provide valuable examples of the development of language.

learning sign language provide natural case studies. For instance, a profoundly deaf boy known as Simon was raised by parents who had learned sign language in their teens. Their gestural speech was crude and inconsistent. Although Simon had seen only their sign language, his own was elegant and grammatical, following subtle rules that his parents did not know. "Complex language is universal," writes Stephen Pinker, "because *children actually reinvent it,* generation after generation."

» Adolescence

Language is fairly mature at age seven, and Piaget's stages of development trail off at twelve-plus, but physical and mental changes continue throughout the life span. Adolescence marks a period of considerable

> **"Education is not the filling of a bucket, but the lighting of a fire."**
>
> POET WILLIAM BUTLER YEATS

emotional and social turmoil. Teens develop greater abstract reasoning and apply it to the world around them, discovering fallacies and injustice. They examine their own identities and test out roles. How do I fit into my family? Where do I fit in at school, or with my friends? How am I special? How am I the same? Peers become more influential as parents lose sway. However, most adolescents remain deeply attached to their parents, if unwilling to show it. A survey of thousands of adolescents across ten countries found that most of them say they like their mother and fathers. Most teens end up adopting their parents' religious faith and political views. The lesson for parents: If you're worried about your teenager's rebellious views, wait a few years. They probably haven't traveled too far from you after all.

As we age, we become choosier about our friends and social lives.

Much to the frustration of adults, adolescence is often marked by risky behavior and impulsive decisions. Teens can blame their brains for that. The human brain continues to develop into the early twenties. In the teen years, areas that had vigorously built up neuronal connections begin to be weeded

Adolescents begin to investigate their own identities even as they hide their inner selves from their parents.

out. Underused connections are pruned, making the brain a more efficient organ. Meanwhile, emotional centers in the limbic system reach maturity well before the frontal lobes that govern rational decision-making. There's a good reason why mortality rates and crime statistics soar in the late teen years. The cerebral centers of reason aren't fully online until about age 25. Passions run high, but self-regulation lags.

GROWING THROUGH ADULTHOOD

Inner growth doesn't stop at 25. Although most attention has been paid to the dramatic changes of childhood, psychologists are now realizing that adults, too, proceed through predictable life stages. How they weather those stages, and the mental attitudes they bring to them, will shape their ultimate well-being in life.

It's tough to study people across the life span. By definition, that's a project lasting 75 years or more. Harvard researchers led by George Vaillant managed it, however, in their pathbreaking Harvard Study of Adult Development (also known as the Grant Study). The study followed 268 Harvard students (all men) from the classes of 1939 through 1944. Vaillant and company collected reams of information on the men, ranging from physical and mental health examinations to life histories. Then they tracked them through the years. The study continues to this day.

Not surprisingly, given the privileged position from which they began, many of the men in the study achieved remarkable success. Most identities are still concealed, but they included a best-selling novelist, a presidential cabinet member, and one president: John F. Kennedy, whose identity is known because his file is the only one sealed until 2040. Yet other men suffered as life went on. Like the population in general, many struggled with mental illness; alcoholism derailed other lives in midstream.

Vaillant's biggest takeaway from the study was not the ways

Growth does not end with childhood. Adults continue to mature throughout their lives.

in which lives veered off course, but the ways lives proceed through health and well-being. He identified five stages of adult development:

- **Intimacy:** the capacity to live with another person in an interdependent, committed relationship
- **Career consolidation:** commitment to a career, compensation, contentment, and competence
- **Generativity:** the assumption of responsibility for the growth of others
- **Guardianship, or caretaking:** charitable work, volunteering, or the curating of cultural riches for future generations
- **Integrity:** the ability to come to terms with the past and future in the face of dwindling days.

People who reach the generativity stage of life are healthier in old age.

Men who were able to grow through these stages fared better than those who stalled out. By 2011, only 4 of 31 men who failed to mature past the intimacy stage were still alive. Fifty of the 128 who reached generativity survived; those who died at that stage still lived an average of eight years longer than the men who stopped at career consolidation. Furthermore, those who stopped at the career stage shared a lifelong inability to deal with anger.

The strongest predictor of lifelong success: love. A warm family life in childhood and strong relationships in adulthood were directly tied to physical, mental, and financial health. The Grant Study confirmed what other research has shown. Social support gives people the strength to take reasonable risks, reach for meaning, and connect with others in both personal and professional life.

The strongest predictor of lifelong success? LOVE

Having a rough start in life doesn't block a person from happiness, however. Although men with difficult childhoods were less likely to reach the generativity stage of life, those who overcame adversity to find purpose were even more likely to cope well with aging. Such was the case with the man identified as "Dr. Camille" in the Grant Study. Coming to Harvard from a cold and paranoid family, Camille was a troubled student, reporting frequently to the infirmary. The college physician noted that "this boy is turning into a regular psychoneurotic." After graduation, Camille attempted suicide. Yet as the years went on, he showed more signs of stability, acquiring a medical practice, marrying, and perhaps most crucially, having children. Although he still had times of struggle, Camille later described the experience of giving back as a physician, and of raising children, as turning points in what became a satisfying life.

"Before there were dysfunctional families," Camille said, "I came from one. My professional life hasn't been disappointing—far from it—but the truly gratifying unfolding has been into the person I've slowly become: comfortable, joyful, connected, and effective. Since it wasn't widely available then, I hadn't read that children's classic, *The Velveteen Rabbit,* which tells how connectedness is something we

"I'm not a teacher, but an awakener."

POET ROBERT FROST

must let happen to us, and then we become solid and whole."

Dr. Camille's story illustrates a central finding in current studies of adult growth and well-being. People with a positive attitude, social connections, and a sense of meaning fare much better than others as they age.

Our identities are a work in progress throughout our lives.

» Learning What Matters

Most young people, understandably, see time stretching out before them in a wide expanse and set their goals accordingly. There's time to experiment, to invest in learning new information and skills, to try out new professions and meet new people. As we age, our time horizon approaches and our goals narrow down. We become selective about what we do and with whom we do it. Older people gradually invest more effort into a few close relationships and into tasks they find meaningful. And contrary to the cliché of the grumpy old man, they become increasingly positive—a trait that promotes long-term mental and physical health.

Experiments show, for

Older adults remember positive images better than negative images.

instance, that memory in older adults is oriented toward emotional significance. In a study in which younger and older people had to remember the gender of a speaker as well as the information that person conveyed, younger participants fared better in remembering whether a man or woman had spoken. However, older folks remembered the emotional content of the speech just as well as young ones. Their overall memory had not declined so much as it had changed in its priorities.

Similarly, older people show a memory bias toward positive information. Younger adults weigh negative material more heavily than positive material and spend more mental time processing it. This reverses with age. In one study, experimenters showed positive, neutral, and negative images on a computer to three groups of people: young, middle-aged, and older adults. The older the participants, the greater the ratio of positive to negative images recalled. When the experiment was repeated while the participants' underwent fMRI scans, older adults showed greater activation in their amygdalae, the brain's emotional center, when they saw positive images as opposed to negative ones.

As we age, we also become choosier about our friends and our social lives. Adolescents and young adults are explorers. They're trying to understand how the world works, how other people think, and where they fit in. Typically, they value novelty and new relationships as they break away from their parents. Older adults, who have lived independently and formed new friendships and families, now consolidate their gains. They're less motivated to pursue meaningless social contact—the stranger at the party, the stimulation of a new neighborhood—and instead turn to long familiar

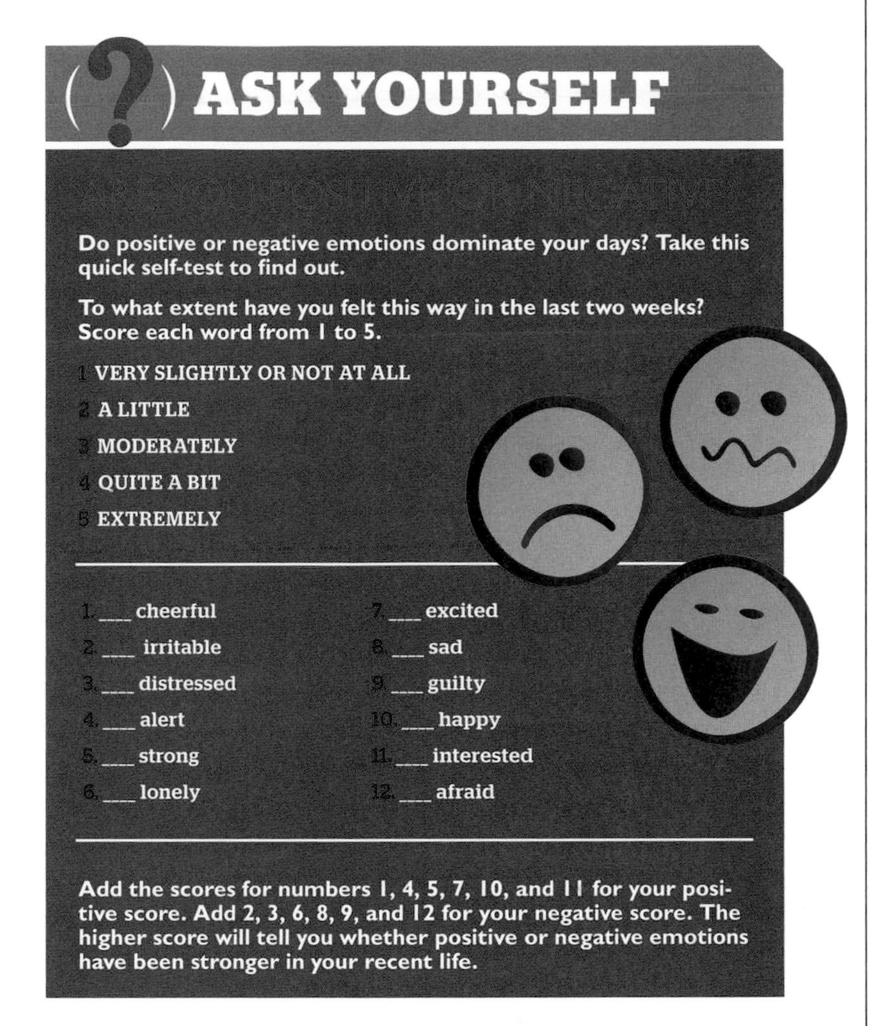

ASK YOURSELF

ARE YOU POSITIVE OR NEGATIVE?

Do positive or negative emotions dominate your days? Take this quick self-test to find out.

To what extent have you felt this way in the last two weeks? Score each word from 1 to 5.

1 VERY SLIGHTLY OR NOT AT ALL
2 A LITTLE
3 MODERATELY
4 QUITE A BIT
5 EXTREMELY

1. ___ cheerful
2. ___ irritable
3. ___ distressed
4. ___ alert
5. ___ strong
6. ___ lonely
7. ___ excited
8. ___ sad
9. ___ guilty
10. ___ happy
11. ___ interested
12. ___ afraid

Add the scores for numbers 1, 4, 5, 7, 10, and 11 for your positive score. Add 2, 3, 6, 8, 9, and 12 for your negative score. The higher score will tell you whether positive or negative emotions have been stronger in your recent life.

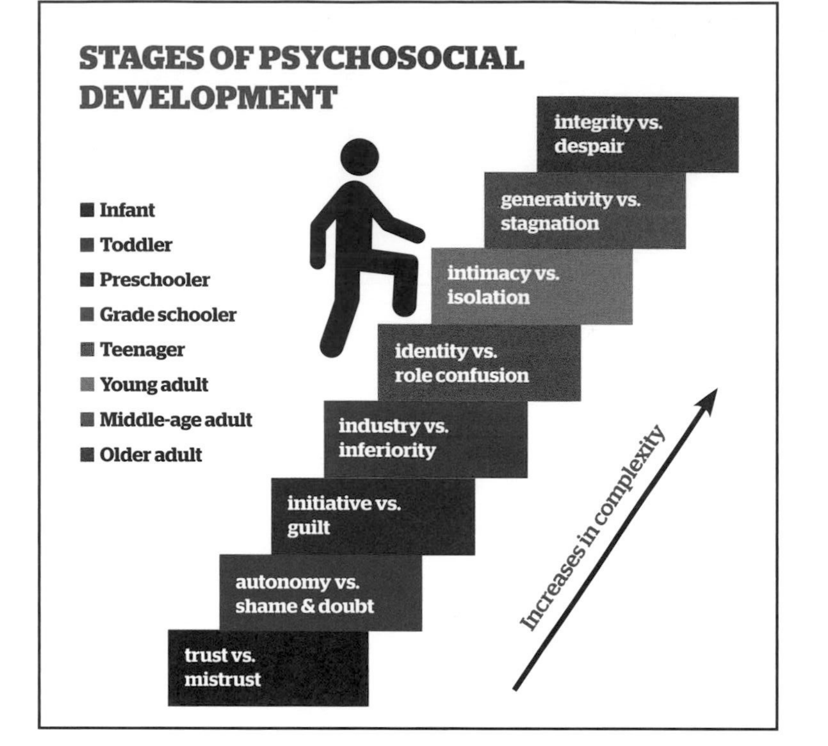

family and friends, valuing the emotional support they receive from those closest to them.

Feeling positive doesn't simply avoid the unpleasantness of sad, crabby, rotten emotions. Over and over again, positive emotions have been linked to better health and improved longevity around the world. In a community of 65- to 99-year-old Mexican Americans, for example, those with more positive emotions at the start of a study were half as likely to die during the two-year follow-up than those with low ratings for positivity. Similarly, in an elderly Swedish community, the answer to "How happy are you with life in general?" predicted mortality—even after controlling for other factors, such as health and ability to perform daily tasks.

Positive emotions influence adult health along several basic pathways:

- **They inspire people to adopt better daily health habits.** Upbeat people are more likely to start and continue with healthy practices such as eating nutritious diets, exercising regularly, and sleeping soundly.
- **They reduce the effects of stress hormones.** The body has a natural fight-or-flight reaction when faced with dangerous situations: The hypothalamus, a tiny portion of the brain just above the brainstem, prompts the adrenal glands to release cortisol. This hormone increases sugars in the bloodstream, suppresses the digestive system, and changes how the immune system responds. These are terrific adaptations when you're being chased by a grizzly bear, but not so useful when stressed by the events of daily life, such as a tough work deadline or simply being stuck in a traffic jam. A steady flow of cortisol in the bloodstream is bad for your overall health. However, positive emotions have been shown to reduce daily cortisol output in people on both working and nonworking days, independent

"We don't stop playing because we grow old; **we grow old because we stop playing."**

PLAYWRIGHT GEORGE BERNARD SHAW

of other factors such as age, sex, or habits such as smoking.

• **They protect against the ravages of age.** Pain, inflammation, and general disability increase with age, but positive emotion appears to dampen these stressors. Positive older people are less likely to be faced by a wide range of conditions ranging from the common cold to stroke, accidents, and rehospitalization after heart attacks.

• **They help you recover.** Stressful events boost our heart rate, but positive emotions can help the heart return to normal. In a study of 170 people given a stressful task (preparing a speech on why they are a good friend), the subjects were then shown one of four films that would bring out the feelings of contentment, amusement, neutrality, or sadness. The contentment- and amusement-inducing films brought their hearts back to normal faster than the neutral or sad movies.

»Well-Being in Adult Life

Positive emotions are obviously key to having a healthy adult

A feeling of contentment, particularly with older people, literally eases the heart.

life. However, overall well-being involves more than just physical health. What are the most important factors to living the good life, psychologically, as adults?

University of Wisconsin psychologist Carol Ryff, who has extensively studied optimal aging, identified six main components to mature well-being:

• **Self-acceptance.** You have a positive attitude toward yourself; you acknowledge and accept multiple aspects of yourself, including the good and bad qualities.

• **Positive relations** with others. You have satisfying, trusting relationships with others; you are concerned about the welfare of others; you understand the give-and-take of human relationships.

• **Autonomy.** You are independent, able to resist social pressures to think and act in a certain way.

• **Environmental mastery**. You feel in control of your environment. You take advantage of your opportunities effectively.

• **Purpose in life.** You have goals and a sense of direction. You feel there is meaning to your present and past life.

• **Personal growth.** You believe that you are continuously developing, growing, and expanding. You have the sense of realizing your potential.

It is a lucky and content person who has achieved all of these states of being in adulthood. Typically, feelings of environmental mastery and autonomy increase with age. The teen years and young adulthood are times of turbulence, when we are subject to the control of our parents, teachers, and bosses. As we move toward middle age, we gain more control over our lives. On the other hand, personal growth and purpose in life tend to diminish with old age. Many middle-aged people feel they've accomplished as much as they can. Perhaps their children, who gave them so much direction in life, have moved away to lives of their own. At this stage, people need to actively seek out new goals to keep moving forward.

Turning eighty? Come on, you're still a youngster: I'll turn one hundred and twenty next month...

Good relationships with others are one factor in well-being.

Two key events affect our eventual psychological well-being: parenting and the decline of physical health. Children give parents a sense of meaning and accomplishment, for good or ill. Research on middle-aged adults shows that their perception of how their children turned out accounts for up to 30 percent of their sense of environmental mastery, purpose in life, and self-acceptance, as well as their feelings of depression. Physical health is an obvious factor in well-being, but it comes with a surprising twist. How we feel about our own health is, to some degree, competitive. We measure our health against that of others like us, and if we believe we're doing better than our cohort, our sense of well-being improves.

We're proud of our family.

THE STORY OF YOUR LIFE

Each one of us is living his own autobiography—an evolving story with significant chapters, key scenes, turning points, main characters, and lessons. We organize our experiences into stories that explain who we are, how we got here, and where we might be going. We have a narrative identity.

Researchers have begun to study personal narratives as a complement to more traditional measures of personality such as traits or motives. Life stories tell us much about how people see

“The most important thing is to enjoy your life—to be happy—it's all that matters.”

ACTRESS AUDREY HEPBURN

themselves and how they fit themselves into the larger narratives of their own societies.

The episodes in our life narratives typically touch on common themes. Those include:

• **Agency:** Ways in which we've changed our lives or achieved our goals
• **Communion:** Connection with other people—love, caring, a sense of belonging
• **Redemption:** Times when we've overcome adversity to reach a good outcome
• **Contamination:** Scenes in which a good event turns unexpectedly bad
• **Meaning making:** Times when we've learned from an event
• **Exploratory processing:** The extent to which we've explored our own lives through our story
• **Positive resolution:** Ways in which we've resolved conflicts in our life stories to come to a satisfying conclusion.

We start to write our own stories early in life, but the narrative keeps changing over time. By the age of two, toddlers begin to collect and tell stories about events that happened to them. Elementary-school-age children can tell coherent stories about their lives with a classic beginning, middle, and end. In adolescence, a sense of cause and effect enters the narrative: "This event led to that one, and I behave a certain way because of my life experiences." At this

Each one of us has an evolving narrative for our own life.

point, a person has gained a narrative identity, although that identity is a work in progress for the rest of her life.

Narratives change over time not only because of growing maturity, but also because memories gain or lose significance as time goes by. Our perspectives change as we reconstruct the past. College students who were asked to describe ten key events in their lives listed only 22 percent of those same events when asked again three years later. Moreover, memory

> **The brain continues to develop into the early twenties.**

is notoriously inaccurate. Even "flashbulb" memories—vivid, emotionally charged recollections of events such as a national tragedy—are unreliable. For instance, researchers asked students to write down their memories of the events of September 11, 2001, on the following day and one, six, and thirty-two weeks later. Their recollections faded, and inconsistent details entered, at the same rate as their memories of unremarkable events. The only difference: The students were much more confident that they remembered the flashbulb event clearly.

As we tell our life stories, most of our narratives hinge on turning points, key life events large or small. They might include receiving a crucial

word of praise from a teacher, experiencing a parental divorce, getting into college, coming out as gay, surviving a serious illness, or having a first child. (Indeed, researchers have discovered that reminiscences include a larger number of events from the ages of 10 to 30 than any other period. This "reminiscence bump" probably occurs because that passage contains more crucial life decisions and social contacts than any other time.) Our natural tendency as storytellers is to connect these dots. We integrate these moments into a narrative with consistent themes and lessons.

How we handle negative events in our stories turns out to be particularly significant. We tend to find similarities among bad moments and to come up with explanations for them. The ability to process crises and find positive meaning in them is linked to psychological maturity, resilience, and greater well-being in the long run. Researchers have also found that people who distance themselves from a past problem by viewing themselves, essentially, in the third person, are better able to process the issue and less likely to become upset. Framing an event as a story with a separate protagonist apparently helps us make sense of problems and cope with them.

Our life stories are important not so much for their autobiographical details—indeed, our memories are typically selective and skewed—but for the meaning we give them. We cobble together a personal narrative in order to give our lives a consistent theme and purpose.

Like all stories, our personal histories are made to be told. We can recite them to ourselves, but they gain power and significance when we tell them to others and take in their reactions. People retell their stories of memorable events soon after the event takes place and repeatedly afterward, and they tailor their narratives to the expectations of their listeners. Researchers have found that a listener's response has a strong effect on the storyteller. A storytelling session lasted only half as long when the listener was inattentive, and the storyteller regarded her own story as less significant if the listener was distracted. Attentive listeners, even hostile ones, boosted a storyteller's

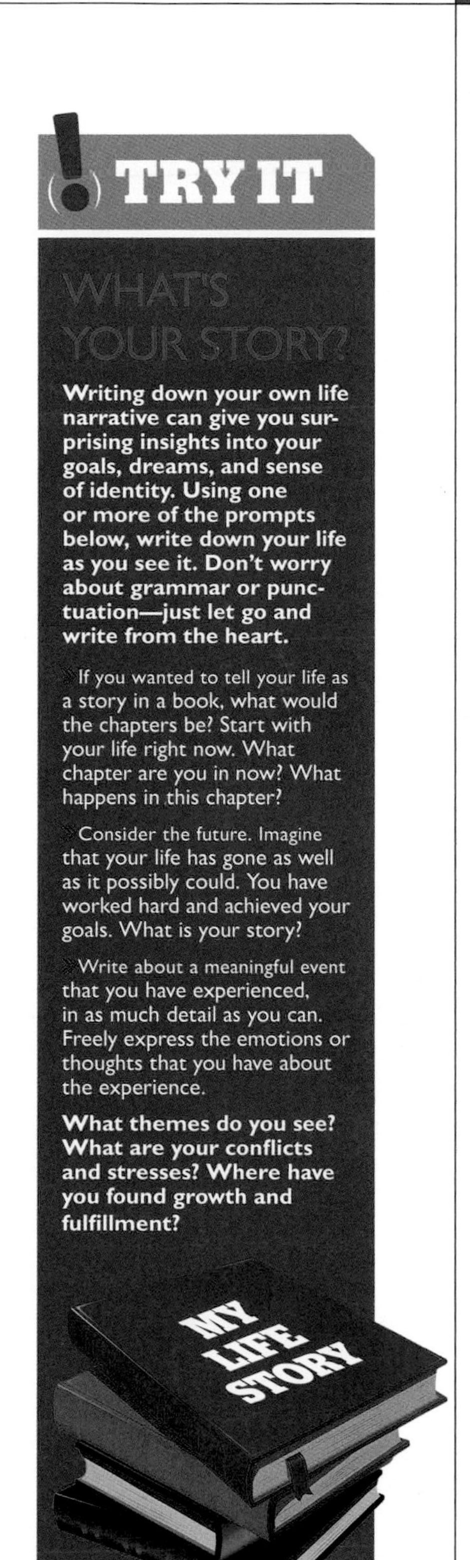

“Everyone is necessarily the hero of his own life.”

NOVELIST JOHN BARTH

confidence that the narration was meaningful.

Parents will be unsurprised to learn that while young adolescents like to share stories with family, as they get older they prefer their peers as an audience. With romantic partners, the more the two agree on the meaning of a memory, the more the teller is likely to retain it over time.

»Cultural Contexts

Each life may be unique, but we see our own life stories in the context of our culture. As Northwestern University psychologist Dan McAdams notes, “Stories live in culture. They are born, they grow, they proliferate, and they eventually die according to the norms, rules, and traditions that prevail in a given society.”

Stories follow certain rules with which we are familiar almost from birth. In modern Western culture a protagonist will be shaped by his family, grow through certain formative experiences, experience insights and epiphanies that direct his later course, and progress toward success or failure. Huck Finn escapes his brutal father, learns from his encounters with various con artists and hypocrites along the Mississippi River, and realizes that he must be true to his own moral compass. Jane Eyre, an orphan, perseveres through a difficult childhood, stands up to the domineering Mr. Rochester, realizes her own abilities, and returns to Rochester on her own terms.

In telling our own stories, Westerners tend to follow these plotlines as well. Typical life stories will follow the lines of classic tales. If you think about your own life story, you probably see it in terms of a classic narrative: My birth into a certain kind of family, key childhood experiences, turning points, setbacks overcome, eventual success. What we remember of our lives is shaped by what we expect to remember.

Because our vision of our own life is shaped by cultural expectations, people from different cultures will see their own lives in different ways. North American adults, for instance, will recount earlier and more detailed memories of childhood

People from different cultures identify different high points in their life stories.

than East Asian counterparts. When Americans from a European background were compared to Chinese adults, the Americans had more memories of one-time events and their roles and emotional reactions to those events. Chinese adults had more memories of social and historical events and placed more emphasis on their social interactions.

A study that compared typical Dutch and Japanese "life scripts" found five events in common: marriage, the first full-time job, having children, beginning school, and parental deaths. However, the Japanese scripts included some unique milestones as well: high school entrance exams, major achievements, the *seijin-shiki* (coming of age) ceremony, the *shichi-go-san* festival for children, and reaching adulthood. The Dutch scripts had three major events lacking in the Japanese stories: leaving home, the first sexual experience, and the death of grandparents.

> **Our vision of our own life is shaped by cultural expectations**

If our vision of our lives is shaped by the dominant culture, does that mean that minority or marginalized groups are left out? Do women or ethnic minorities lack a template for their lives? Opinions differ. Some researchers believe that oppressed groups have their

The role of grandparents looms larger in some cultures than in others.

own characteristic counternarratives of living outside of the mainstream culture.

GIVING BACK

According to his instructions, Thomas Jefferson's gravestone reads:

Here was buried
Thomas Jefferson
Author of the Declaration
of American Independence
of the Statute of Virginia for
religious freedom
and Father of the University
of Virginia

What's missing? A little thing like being president of the United States. Jefferson's epitaph reflected the accomplishments of which he was most proud. The Declaration, the statute, and the university were his legacy to the world. Few people can match Jefferson's achievements, but with age many begin to reflect upon how they can give back to society before they're gone. Vaillant called this stage in life "generativity." It's a particularly important, and healthy, part of an individual's life trajectory, and a factor in adult well-being. Younger and older adults with substantial generativity strivings show greater recovery in the aftermath of stress, failures, and other life problems. Cultivating generativity can help you weather old age.

Giving back to others is an important part of well-being in adults.

Generativity takes many forms in daily life. It can encompass creativity: writing a book, creating art, making music. It may involve giving back to the community through volunteer work. Families are fertile grounds for generativity, whether it's helping children with schoolwork or sports, or taking care of an elderly parent. Parents are more likely to be generative, and those who score highest on a generativity scale are more involved in their children's lives, from attending parent-teacher meetings to volunteering in schools. On a grander scale, generativity may be directed toward a kind

“Life itself is the most wonderful fairytale of all.”

AUTHOR HANS CHRISTIAN ANDERSEN

of immortality. People want their lives to have meaning after they're gone. They long to have a lasting influence on their company, their community, or even the world.

Narratives of redemption are particularly common among people with strong generative motives, as well as among addicts in recovery and prisoners attempting to reform. These individuals might describe a turning point in which they gave up alcohol or drugs and instead began to rebuild their lives and help others do the same. Those who have suffered through the death of a spouse may later say they have gained strength or compassion from the experience. Generative people are more likely to see these tough episodes as times when negative eventually turned to positive. (The opposite of redemption, contamination, is a point in which a positive life takes a turn for the worse. Generative folks describe fewer of these moments.)

A sense of redemption is a powerful force for well-being—even more powerful than having a happy life story. People who believe they have turned bad into good, that they have overcome obstacles, feel better than those who report that life has been pretty good all along. As author bell hooks, who overcame a tough childhood to become a teacher and activist, has written, "unnecessary and unchosen suffering wounds us but need not scar us for life. It does mark us. What we allow the mark of our suffering to become is in our own hands."

From infancy to old age, we all follow certain scripts. We all grow in predictable ways. We acquire speech and social skills and mature capabilities. We make mistakes and we learn from them. This pattern of development is dictated to some extent by universal human biology, but biology is not the whole story. As we age, we can reflect on the people we've become, on our own life story, and find a meaning in that story that helps us understand our impact on the world.

The MIND &

WORLD

None of us lives in isolation. The most remote hermit on the farthest mountaintop has to interact with his surroundings. He has to make choices, take action, and solve problems. Most of us live in a far more communal world, where our personalities and abilities are defined by the ways in which we navigate our social environment. In the next three chapters, we'll consider what drives us as we relate to each other one-on-one and in groups. We'll encounter some surprising findings about how we reason and create. And we'll learn how we can harness our capacity for self-control to reach our goals.

I couldn't do this without my friends to support me.
We're all in this together.

» CHAPTER FOUR «

SOCIAL ANIMALS

Being excluded is painful. Literally painful. Consider, for instance, the Cyberball experiment. In 2003, researchers reported on the reactions of players to a simple online, virtual ball-tossing game. The experimenters told participants that they would be playing the Cyberball game with two others on remote computers. Unknown to the participants, the other players were just computer simulations. In the game's first round, the players tossed the ball around freely. In a later round, the other players stopped throwing the ball to the participants, excluding them, no matter what they tried. The rejected participants felt hurt, and not just in a metaphorical way. Scans of their brains showed greater activity in regions associated with the emotion of actual, physical pain—the same sort of pain you might feel if you smashed your thumb with a hammer.

"Without friends," wrote Aristotle, "no one would choose to live, though he had all other goods." Humans have a fundamental need to relate to others. It's a desire that drives them throughout their lives. Infants are attuned to their mothers from the moment of birth. Teens turn to their peers for validation and support. Relationships and connectedness are keys to well-being in older adults, to the point of boosting longevity. We are profoundly social animals, deeply influenced by our ties to others. Our relationships shape our sense of our own identity not only one-on-one with close partners, but in larger groups that mold our beliefs and attitudes. And in a broader sense, our personalities, decisions, motivations, even our intellectual capacities, are all designed to help us navigate the social environment. Our social needs may be based in our biological inheritance, but they are expressed in all the ways that we react and interact with the outside world.

“Society is not the mere sum of individuals, but the system formed by their association.”

SOCIOLOGIST ÉMILE DURKHEIM

BONDING AND REJECTION

Our social selves are deeply comforted by one-on-one bonds. To truly feel connected to another person, we need to experience a stable and lasting relationship. In such a bond, we interact with a close friend frequently and pleasantly, without hiding our true selves. Fleeting contacts with a variety of strangers are not emotionally satisfying. We need to feel that the other person knows us and has our well-being at heart. People who have these kinds of close relationships are demonstrably happier and healthier than others.

Conversely, when we are cut off from close relationships, shunned, or rejected, the experience can be devastating. Think about times you've gone through a romantic breakup, or a family spat in which a loved one cut you off. Even as small children, some of our worst moments come when our best friends move on to others, or when a group of kids keeps us on the outside. These moments are inevitably described as painful. In fact, a whole vocabulary of pain describes emotional breakups: Our feelings are hurt, our hearts are broken, we feel the sting of rejection. As the Cyberball experiment shows, this is no accident. Feelings of rejection do indeed register as physical pain in the brain.

The anterior cingulate cortex (ACC), located under the brain's frontal lobes, and the anterior insula, part of the emotion-processing limbic system, are both closely connected to communicating information about pain and emotion. In fact, research suggests that pain registers in the brain in two ways: as sensory information about the location of the pain, and as emotional information about how nasty that sensation is. When an infant cries in distress at being separated from its mother, that wail is prompted by the ACC. In cases where surgeons have removed portions of the ACC, patients report registering the sensation of pain without experiencing it as unpleasant. Rejected Cyberball players didn't feel physical pain, but they felt the distress of their exclusion just as they would feel the distress of taking a punch. The more rejected they felt, the greater the pain.

For those inclined to rejection, simply looking at paintings of isolation, such as this one by Edward Hopper, can be painful.

Some people seem to be more attuned than others to rejection and more sensitive to the pain it causes. For these folks, looking at paintings of lonely, isolated people (such as those by Edward Hopper) lights up the emotional pain centers in the brain, as does viewing photos of people with disapproving expressions.

Even in retrospect, people feel actual pain when contemplating social loss. The ACC region is more active when people view images of loved ones who have died recently, versus images of strangers. Similarly, women who have lost unborn babies register pain in the ACC when they see pictures of smiling baby faces.

So, if emotional pain overlaps with physical pain, can you take an aspirin to make it go away? Surprisingly, the answer is yes—to some extent. Studies have shown that people who took Tylenol each day for three weeks reported much lower levels of daily hurt feelings, starting on day nine, than those who took a placebo. The same results were seen when the rejected Cyberball players took a daily Tylenol.

Tylenol treats emotional pain as well as physical.

Cultural context also shapes how sensitive we are to being rejected or ostracized. Take, for instance, farmers and herders. In a study conducted in Turkey, researchers questioned both farmers, who worked independently and sold their

FOCUS

SHUNNING

Such is the power of being excluded that formal shunning has been used for millennia to discipline and punish errant members of social groups. Ancient Athens gave birth to the word ostracism through its practice of *ostrakophoria.* At midwinter meetings, citizens would meet and cast votes on shards of pottery *(ostraka)* in order to expel prominent citizens considered to be threats.

Several religious groups practice shunning, perhaps most famously some churches among the Old Order Amish and the Mennonites. Baptized members who are believed to have broken their vows are cast out of the community. Other Amish will not eat or do business with them until they have repented. More drastic is the traditional Balinese practice of *kapesekang,* in which local and political disputes can lead to a sentence of complete ostracism. The offender's neighbors will not look at or acknowledge him; he and his family are banned from temples and cannot use schools or health clinics, because no one will speak to them. When the shunned person dies, he faces the ultimate ostracism: His body cannot be buried in the community cemetery.

produce directly to stores, and herders, who regularly had to negotiate one-on-one with strangers to sell their cattle. Both the farmers and the herders reacted with similar amounts of pain to the experience of being shunned by those close to them. When it came to ostracism by strangers, though, the two groups differed. The farmers, who didn't rely upon dealing with strangers for a living, were less disturbed by getting the cold shoulder from unknown people. Herders, who need that interaction, were more strongly affected.

So when you feel as though your heart is broken, or when a cutting remark truly feels cutting—take it seriously. Your pain is real.

From a modern viewpoint, it seems odd that seemingly harmless social slights should arouse deeply rooted physical responses. Seen in an evolutionary perspective, however, the pain of exclusion is reasonable. Humans are completely dependent upon parents and caregivers for food and protection during their long period of immaturity. For most mammals, to be cut off from caretakers during childhood means death. In early human history, adults, too, depended upon their peers for protection. Social groups were small and often in peril, with the members relying upon each other in order to eat, raise young, and defend against predators. Being cast out from the group was a calamity. As UCLA psychologist Naomi Eisenberger writes, "Given that being socially disconnected is so detrimental to survival, it has been suggested that in the course of our evolutionary history the social attachment system—which ensures social closeness—piggybacked onto the physical-pain system, borrowing the pain signal to cue instances of social separation." Prompted by these unpleasant feelings, people hasten to band together and avoid the dangers of a solitary life.

»The Costs (and Benefits) of Rejection

The pain of hurt feelings following rejection is often accompanied by an unwanted emotional partner: anxiety. Being excluded makes most people tense and anxious as they sense that trouble has invaded their lives. Again, this makes evolutionary sense. This unpleasant emotion warns us that we're in danger of being cut off from the support of our social group. To dispel the anxious feeling, we'll strive to reconnect.

> **Desired inclusion in social groups is a factor of evolution.**

Anxiety can arise even before the actual rejection, triggered by warning signs such as criticism. Critical remarks from a family member or boss, or comments from a loved one that you are losing your looks, for instance, are signals that exclusion may be on the way. We may be angered by the criticism, or we may disagree, but the anxiety provoked by the comments arises nonetheless. Embarrassment and guilt serve similar functions to anxiety, but in retrospect. Our guilty feelings help us to avoid future social blunders and to repair our relationships.

To a certain degree, anxiety,

Being excluded from social groups can lead to anxiety.

TRY IT

COPING WITH REJECTION

Everyone has been criticized or rejected from time to time, and everyone hates it. However, there are ways to cope and come out of the experience a stronger person. Psychologist Todd B. Kashdan suggests a few useful strategies:

1. Look at the person doling out the criticism. Is the critic a creator or a destroyer? Vitriolic comments are sticky, but listen instead to creators who try to build on what you created.
2. If your work is being rejected, create psychological space between you, the creator, and the work, your product. Learn to create distance between the thoughts and the thinker, the feelings and the feeler.
3. Take your unwanted thoughts and repeat them to yourself slowly. Train yourself to see them for what they are—just words, just sensations, not something that has control over what you do next.
4. Understand the difference between productive and unproductive distress following rejection. Ask, "Is there anything I can do today that would solve my concern?" If the answer is nothing, acknowledge this and move on to something productive.
5. Invest in your tribe. Share your vulnerabilities and imagination with people close to you. Be kind even in a cruel world.

embarrassment, and guilt are valuable, because they protect us from the dangers of social exclusion. Taken further, however, the crippling disorder of social anxiety harms the person who experiences it. Social anxiety is one of the most common psychological disorders (see chapter 7, page 203). People who suffer from it are unusually afraid of humiliation or rejection and accordingly avoid socializing. When forced to interact with others, they focus on the other person, going to great lengths not to reveal their own thoughts or feelings and avoiding controversial subjects, because they fear exposure, embarrassment, and rejection. This can smooth over social interactions, but it keeps other people at arm's length. Socially anxious people may even disparage themselves or downplay their own accomplishments so as not to alienate others. They may avoid public praise or positive evaluation for fear of arousing envy, or because they're afraid they'll create high expectations that they won't be able to reach. Even in success, those with social anxiety attempt to fail, so they won't dislodge their place within the group.

Experiencing rejection may contribute to diabetes and heart disease.

Rejection is bad not only for your mental health, but for your physical health as well. In particular, the experience of rejection seems to stimulate the pathways that lead to inflammation in the body. Persistent inflammation, in turn, can exacerbate conditions such as diabetes and heart disease. These effects have been found in adults giving speeches in front of a hostile crowd and in spouses in conflict. Rejection may be particularly harmful in the adolescent years, when social status and group membership is so emotionally important. In a two-and-a-half-year study of adolescent girls who were prone to depression, markers of inflammation were most pronounced after the girl had experienced an episode of rejection. This was especially true for high-status girls

Girls who see themselves as high on the social ladder suffer more stress than others.

“Wherever there is a man who exercises authority, **there is a man who resists authority.**”

AUTHOR OSCAR WILDE

(who evaluated their own status by picking it out on a ladder, with the low rungs representing unpopular, academically struggling girls, and the high rungs representing, popular, academically successful girls). Why is that? Studies show that high-status people have stronger stress reactions to social threats than those of lower status, possibly because, in evolutionary terms, they have more to lose by being knocked off their lofty rung of the social ladder.

On the other hand, one kind of person may actually thrive on rejection: the proudly independent individual. Author Stephen King impaled one rejection letter after another on a spike in his bedroom as he peddled his novel *Carrie* to publishers, but didn't give up until he sold the book on the 30th try. Studies have found that people who are unusually independent and value their own uniqueness are more creative following an experience of rejection. In tests, they make

Don't hate me because I'm beautiful.

more unusual word associations and produce more imaginative stories and drawings. To these proud outsiders, social rejection may be validating rather than discouraging, energizing rather than depressing.

BRIDGING WORLDS

Building connections to others is clearly vital for our physical and emotional health. How do we do this?

We can start with empathy. Being able to identify with another person's emotions, particularly that person's pain, is central to forming ties with others. Most people, seeing a child bullied on the playground or a friend grimacing after stubbing a toe, will wince in sympathy. We seem to feel what they are feeling. And in fact, we do, quite literally.

Many studies confirm that our brains react to a familiar person's pain as if we were experiencing that pain ourselves. For instance, researchers in Taiwan showed a series of images of hands and feet to a group of young men who were in close relationships. Some images showed the limbs in neutral situations, such as opening a drawer. Others showed the same limbs in painful situations, such as being slammed in the drawer. The experimenters asked the men to envision either a stranger, their loved one, or themselves in each of these scenarios. Functional MRI scans showed greater activation in the brain's pain matrix (our friends the ACC and the anterior insula) when imagining a loved one in pain than when visualizing a stranger in the same situation. In fact, the closer the relationship between the subject and the significant other, the more the subject's brain registered the pain as if it were his own.

Highly empathetic people, such as Mother Teresa, feel other people's pain.

It's not just physical pain that triggers these empathetic reactions. When we see other people experience social pain, such as rejection, the same areas of the brain light up. Remember the Cyberball study (see this chapter, page 109), in which simulated players engaged in a mean game of keep-away? Another group of participants merely watched the interactions. Afterward, they were asked to write an email to the excluded participant. Empathetic people, watching the keep-away, showed activity in the pain-related regions of their brains. Furthermore, those who showed the strongest empathetic reactions wrote the most sympathetic emails. "Dear Adam," wrote one such observer, "While watching your game of Cyberball I noticed you may have felt left out when Erika and Danny were consistently throwing the ball to each other. I just wanted to say I'm sorry that happened and I am sure there is some explanation that has nothing to do with you. You seemed to be a great ball thrower." A less empathetic watcher wrote this: "Hey Ann, Thanks for participating in the game with the other two participants. It was an interesting game, and I hope that you had fun!"

Is this empathetic behavior purely altruistic? Evolutionary explanations would say: not entirely. Pro-social behavior can spread throughout a group, benefiting all. We do for others what we hope they would do for us one day.

"I know exactly how you feel."

»The Spectrum of Empathy

Overactive empathy might be painful, but a complete lack of empathy can be dangerous. Mention the word "psychopath," and the image that comes to mind is of a wild-eyed loner, creepy and disheveled, devoid of normal human emotion except, possibly, hatred. Traditionally, psychology has agreed with this assessment. Theories of psychopathy have held that psychopaths are unable to feel empathy or understand another human's feelings.

But then there was Ted Bundy, the serial killer whose roster of appalling crimes would put him near the top of anyone's list of

> "[Empathy is] the capacity to understand that **every war is both won and lost.**"
>
> AUTHOR BARBARA KINGSOLVER

violent psychopaths. Before his execution in 1989, Bundy confessed to murdering 30 girls and young women and is suspected of killing many more. In addition to kidnapping, raping, and beating his victims, he decapitated at least 12 women, keeping their heads in his apartment. He escaped twice from jail, going on to commit more crimes, before finally being arrested for driving a stolen car.

Not all psychopaths are inclined to violence or criminal behavior.

Bundy could never have succeeded in luring so many women to their deaths if he had been a repellent character. On the contrary, his acquaintances and surviving victims described him as charming, charismatic, and intelligent. The handsome, friendly man seemed the opposite of unfeeling, at one point even working at a suicide crisis center. In fact, it was his understanding of empathy that allowed him to lure in so many kindhearted strangers. He approached many of his victims while wearing a cast on an arm or leg and asking the woman for help carrying items to his car.

Psychopathy has been linked to childhood abuse. Many abusers were themselves abused as children, which may have taught

Serial killer Ted Bundy understood empathy, but didn't feel it.

them to bully others in order to gain power or control. But the condition seems to have a physical basis as well. Researchers studying violent criminals have found that they have stunted fear reflexes as well as a low sensitivity to punishment. Psychologists now believe that psychopaths may also possess a stunted form of empathy. Like pain, empathy seems to have both a cognitive and an affective component. Psychopaths may possess the cognitive part—they understand what empathy is and how it works—without the affective part. They simply don't feel it emotionally. It's a recipe for an effective killer, a predator who can manipulate others without getting bogged down in emotional connection. Ted Bundy certainly fit this picture. He felt no guilt for his crimes. He glibly talked his way out of police interviews and continued to lie, boast, and change his story to the day of his death.

Not all psychopaths are criminals. When law-abiding neuroscientist James Fallon discovered the key signatures of psychopathy in his own brain scan in 2005, he realized that his competitiveness and aggression might have a biological basis. Some psychologists say that up to 4 percent of corporate bosses are also psychopaths, manipulative and callous, but not violent.

ASK YOURSELF

THE MICHELANGELO PHENOMENON

"You make me a better person." This heartfelt declaration is a romantic cliché, but like many clichés, it contains a central truth. Close partners do shape each other, and in the best relationships, they help each other move toward their ideal selves.

Psychologist Caryl Rusbult called this the Michelangelo Phenomenon. Like Michelangelo's uncarved stone, she wrote that "humans, too, possess ideal forms," a collection of dreams, skills, goals, and resources that each person wants to acquire. Supportive partners understand these goals and, over time, actively sculpt their loved ones to help them free their ideal selves from the marble.

This support doesn't have to involve life-changing gestures. It can be as simple as a husband prompting his shy wife to tell a charming story at the dinner table. It does have to reflect what the partner truly wants—not what the "sculptor" prefers instead. Couples who achieve this have been shown to have greater durability through adversity and a greater adjustment to life's challenges.

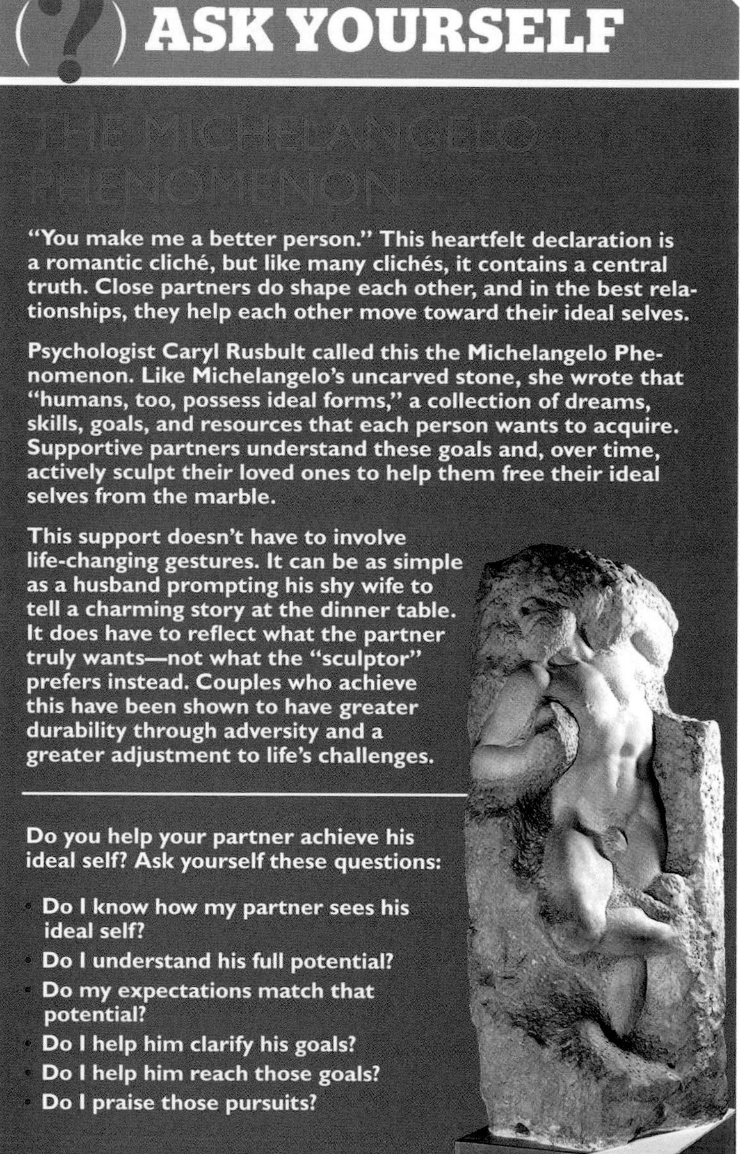

Do you help your partner achieve his ideal self? Ask yourself these questions:

- **Do I know how my partner sees his ideal self?**
- **Do I understand his full potential?**
- **Do my expectations match that potential?**
- **Do I help him clarify his goals?**
- **Do I help him reach those goals?**
- **Do I praise those pursuits?**

» You Get Me

"No one knows me like my best friend." "My boyfriend just gets me." When you feel someone is truly close to you, you believe that person understands you like no one else. When you're interested in getting close to someone else, you want to really know him and comprehend why he does what he does.

Understanding joins empathy as a vital form of social connection. It takes two forms in relationships. The first is knowledge. People want to know others and to be known by them. Even outside of intimate relationships, knowledge is valuable, helping us to grasp other people's motivations and to explain and predict their behavior. We need to know a person before we can decide if he is trustworthy.

The other form of understanding is responsiveness. When you feel your significant other values you, accepts you, and is acting in your best interest, you feel committed and validated. Responsiveness has been shown to be key to fostering feelings of security, intimacy, and trust between partners. Once begun, the process of understanding flows both ways. As one partner helps another reach her goals, that responsiveness encourages the helping partner to open up about his own dreams.

The most important part of mutual understanding is not the knowledge itself, but the sensitivity and emotional support shown. One study found that when it came to knowing a loved one's personality, food preferences, or behaviors, perceived knowledge was more important than actual knowledge. As long as people feel validated, and that their partners have their best interests in mind, it doesn't matter if their loved ones don't truly know their taste in beer.

"You never really understand a person until you consider things from his point of view."

AUTHOR HARPER LEE, *TO KILL A MOCKINGBIRD*

Children thrive when their parents are responsive to their needs.

The "you get me" phenomenon is not confined to romantic partners. This kind of close understanding also underpins good parent-child relations. Children whose parents are responsive to their needs are more securely attached, with better social skills and greater academic achievements.

»Gratitude

When you have a responsive and understanding partner, you almost certainly feel grateful. The emotion of gratitude is another powerful factor in building the bonds between two people. Even outside of a close relationship, gratitude can reinforce positive action within a group and help a troubled person find strength in adversity.

From a practical standpoint, gratitude is simply useful. It's a benefit detector. You're grateful when you get a gift, or receive some sort of benefit that you can link to a specific giver. Gratitude intensifies if you believe the gift was costly to the giver

> **"Gratitude opens the door to...the power, the wisdom, the creativity of the universe"**
>
> AUTHOR DEEPAK CHOPRA

(whether in money, time, or effort); if the gift was personally valuable and well chosen (that bottle of aged Glenlivet is perfect for the Scotch aficionado, not so much for the recovering alcoholic); and when the giver was not obliged to give the gift. A spontaneous, out-of-the-blue present can be the best gift of all. In each case, we're grateful not so much for the gift or benefit itself, but because the act of giving shows that people are thinking about us, that they care about us, and that the world is not such a bad place.

Gratitude is useful also because it's the gift that goes on giving; it reinforces kind gestures. When we thank people for their gifts, including the contribution of their time, support, and wisdom, those people are more likely to be generous again in the future to us and to others. Gratitude also motivates us to make kind gestures to others—to pay it forward. When the person ahead of you in line at Starbucks unexpectedly pays for your coffee, you're motivated to do that for the person behind you. Kindness begets kindness, and we are inspired to be as good as the person who helped us. The emotion of gratitude facilitates a flourishing, healthy society.

However, conveying gratitude is not always easy. Fear, embarrassment, sadness, and anger can accompany our expression of the emotion. Gratitude makes us realize how vulnerable we are, how dependent upon others to navigate the shoals of life. (Men, in particular, have trouble expressing gratitude.) Therapists sometimes urge people to write a letter of gratitude to someone close to them—perhaps a deceased relative or a friend—and to read it aloud. The reading might take place at a tombstone or some other emotionally resonant location. As you might expect, emotions at these times are mixed, love and gratitude mingling with sorrow. Even more fraught are letters (not read aloud) from women to the abusive husbands they have left behind. Although the letters can be therapeutic, as the women reflect upon the ways they became stronger, more independent, and more socially connected following the experience, the gratitude is understandably often buried beneath feelings of sadness and anger.

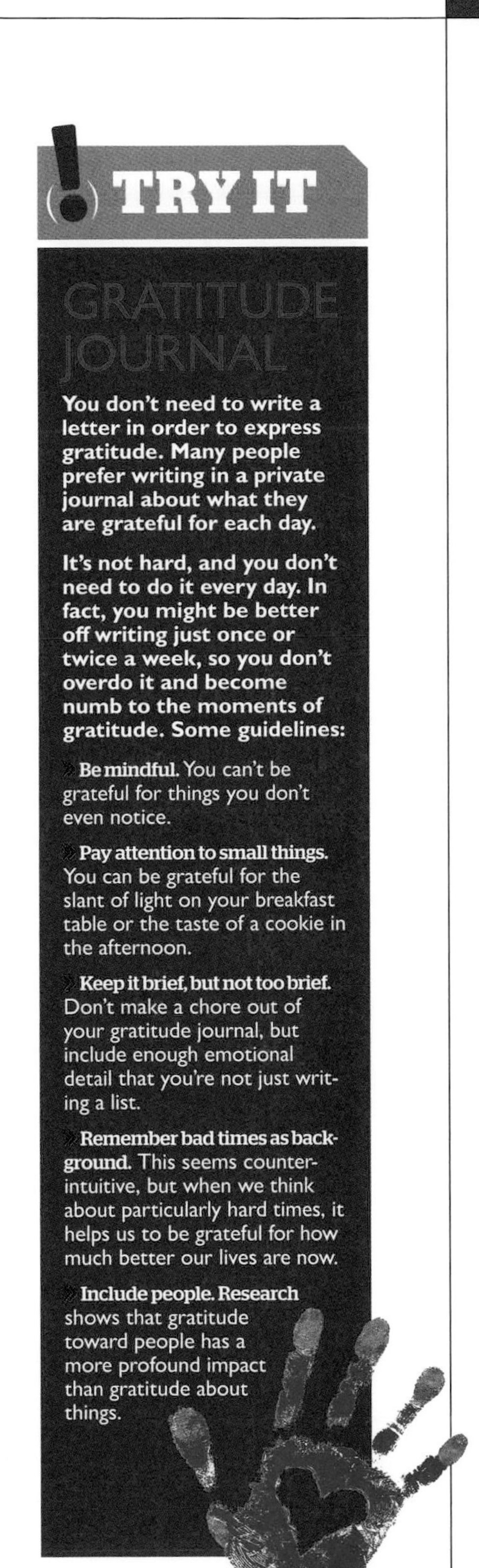

TRY IT

GRATITUDE JOURNAL

You don't need to write a letter in order to express gratitude. Many people prefer writing in a private journal about what they are grateful for each day.

It's not hard, and you don't need to do it every day. In fact, you might be better off writing just once or twice a week, so you don't overdo it and become numb to the moments of gratitude. Some guidelines:

» **Be mindful.** You can't be grateful for things you don't even notice.

» **Pay attention to small things.** You can be grateful for the slant of light on your breakfast table or the taste of a cookie in the afternoon.

» **Keep it brief, but not too brief.** Don't make a chore out of your gratitude journal, but include enough emotional detail that you're not just writing a list.

» **Remember bad times as background.** This seems counterintuitive, but when we think about particularly hard times, it helps us to be grateful for how much better our lives are now.

» **Include people. Research** shows that gratitude toward people has a more profound impact than gratitude about things.

VIOLENCE

FORGIVENESS

LACK OF COMMUNICATION

KINDNESS

NEGATIVE EMOTIONS

OPTIMISM

» Good Intentions, Bad Results

Forgiveness is a positive thing. People who are able to forgive fare better both physically and mentally. They show more satisfaction with life and less overall psychological distress. Even their blood pressure is healthier than that of their less-forgiving counterparts. Relationships typically thrive on forgiveness. Husbands whose wives are more forgiving of a transgression report more effective communications as much as a year later.

And yet—is forgiveness always good? A study of women in domestic violence shelters measured their levels of forgiveness toward their abusive partners. On a sliding scale, the women agreed or disagreed with statements such as "I still hold a grudge against the person in question" and "Although I did not like it, I can accept what happened." Then they were rated on their likelihood of returning to their partner with statements such as "I miss my partner a lot" or "I plan to see my partner again." The more forgiving a woman felt toward the man who hurt her, the more likely she was to return, regardless of other factors such as the severity of the violence. In fact, forgiveness seems to foster violence in abusive relationships. Less forgiving spouses will see a decline in the frequency of aggression toward them over the early years of a marriage. More forgiving partners are punished with stable or increasing levels.

Like medicine, forgiveness is great in the correct amount but poisonous in an overdose. People and their relationships flourish when they are flexible and understanding while still

Practicing gratitude and kindness keeps relationships on the right path.

Forgiveness should be balanced with self-respect in a relationship.

mustering the toughness to recognize and correct a bad situation.

Kindness is another example of a strength that can turn into weakness in the wrong situation. In general, being kind to a partner in times of personal need leads to better satisfaction with the relationship. Even when interacting with strangers, people who perform "random acts of kindness" every day for ten days feel more satisfied with their lives.

Relationships need both kindness and honest confrontation.

On the other hand, kindness can hamper necessary communication. In a study of marriages across four years, wives who were more likely to reject or criticize their spouse during problem-solving discussions (i.e. arguments) actually had better, more stable marriages, as reported by both spouses, at the end of the study. Other research supports the idea that criticism can be a relationship-saver when it pushes people to deal with severe problems. When applied to petty concerns, however, criticism drags down a marriage.

**I know you can't forget,
but you should learn to forgive...**

Kindness is also not a substitute for more concrete forms of support. It works when it's appropriate and responsive. A spouse who is elbow-deep in dirty dishes will not appreciate a kind word nearly so much as a helping hand.

The same rules apply to optimism. Overall, it's a plus. People who expect events to turn out well show greater well-being and stronger relationships over the course of their lives. Optimism decreases stress and depression in young people. However, unbridled optimism can blind people to bad outcomes and bad behavior. Optimists are more likely to persist in gambling, even after losses. Optimistic women who come up with external reasons for their abusive spouse's behavior

We all live in groups. Schoolmates are one kind of natural group.

(he's just stressed out from work) are at greater risk for psychological or physical abuse. In general, spouses who come up with benevolent excuses in the face of severe problems are less likely to solve those problems than those who acknowledge trouble and place some blame on their partner. Nor do we benefit by being optimistic about our own behavior. The negative emotions that accompany bad actions are a spur to action, prompting us to fix what ails us.

GROUP DYNAMICS

All of us live in groups—groups that are part of bigger groups, which in turn make up even larger groups, intersecting with yet other groups in a complex Venn diagram of human relationships. Think about the groups you belong to, and you may be surprised at how many there are. Do you belong to a family? A class in school or a collection of coworkers at the office? A book group, softball team, poker night buddies? Are you linked to "friends" on social media? Twitter followers? A political party, a church or mosque? How about the worldwide community of your coreligionists? Do you identify with a nationality? What about temporary groups—other folks in a ticket line or in a movie theater?

The narrator of Kurt Vonnegut's novel *Cat's Cradle* belongs to a religious group known as Bokononists: "We Bokononists believe that humanity is organized into teams, teams that do God's Will without ever discovering what they are doing. Such a team is called a *karass* by Bokonon." A karass "ignores national, institutional, occupational, familiar, and class boundaries." The opposite of a karass, to Bokononists, is a *granfalloon,* or false team. Examples of this, in Vonnegut's novel, include the General Electric company and "any nation, anywhere."

To those who study social connection, both kinds of groups—the karass and the granfalloon—are legitimate. Two or more people connected by social relationships make up a group. Social psychologists typically separate them into primary and secondary types. Primary groups are long-term, closely knit sets whose members frequently interact face-to-face and have high levels of solidarity. People often become part of a primary group involuntarily, with the family as the prime example. Their close friends and loved ones might also become part of their

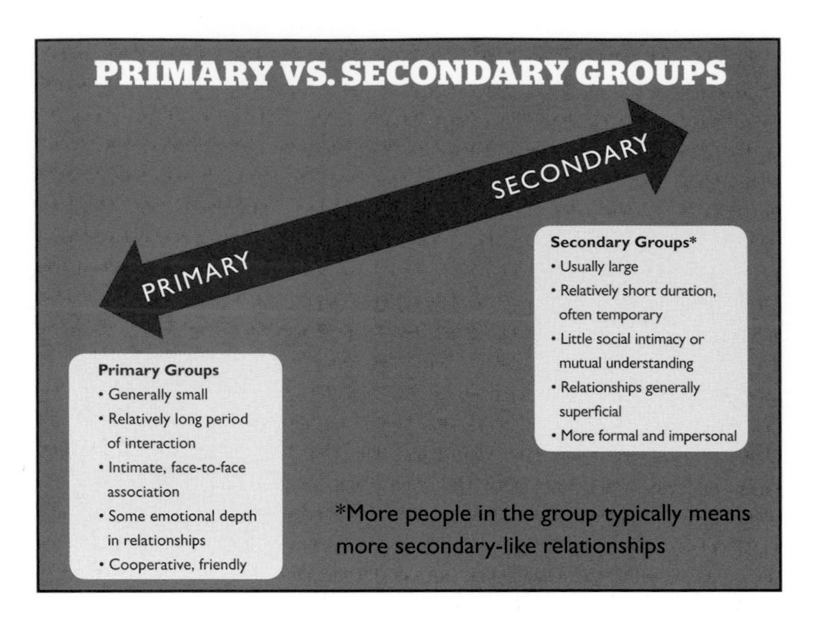

primary group. In early human history, primary groups were the only kind there were. People lived in small tribes or collections of hunters and gatherers, isolated from outside social influences. As societies grew and evolved, however, secondary groups began to form.

Secondary groups come in myriad forms, but in general they are larger and more formally organized, yet less emotionally involving than primary groups.

Secondary groups are crucial, though, in defining an individual's identity and role in society. These collections can range from protesters on a city street to all the employees within a corporation. Researcher Holly Arrow and her colleagues have identified the characteristic ways in which such secondary groups form:

• **Concocted groups** are organized by people or authorities outside the group. The flight crews of an airplane or a

"Society attacks early, when the individual is helpless."

PSYCHOLOGIST B. F. SKINNER

military squad, assembled under the command of an outsider, are examples.

• **Founded groups** are begun by individuals who remain members of the group. Expeditionary teams, book groups, or Internet start-up companies might all be founded groups.

• **Circumstantial groups** emerge in response to temporary outside forces. Airline passengers stranded in a snowy airport together or a mob breaking store windows is a circumstantial group.

• **Self-organizing groups** may also arise in response to circumstance, but their members gradually align and cooperate with each other. Drivers waiting to exit from a crowded parking structure and teenagers at a party, deciding to turn up the music and dance, are groups whose actions become increasingly coordinated through minute social adjustments.

Secondary groups are like organisms. They have a predictable life cycle and characteristic dynamics. Ohio State researcher Bruce Tuckman identified five stages of group development in a classic study:

Forming. Members become oriented toward one another. They become dependent and focus on a task.

Storming. Conflicts arise as members vie for status and the group debates its goals.

Norming. Conflicts subside as the group becomes more structured and cohesive. Members open up to one another and express their opinions.

Performing. The group turns to its task.

Adjourning. The group has fulfilled its function. Members may feel sad or anxious as the group breaks up.

»Social Influence

Most of us would like to think that we have a mind of our own—that we can think and act independently—but decades of research indicate otherwise. As social animals, we are deeply and rather easily influenced by the norms of our groups and by the power

Groups reach the so-called norming stage when they become more cohesive.

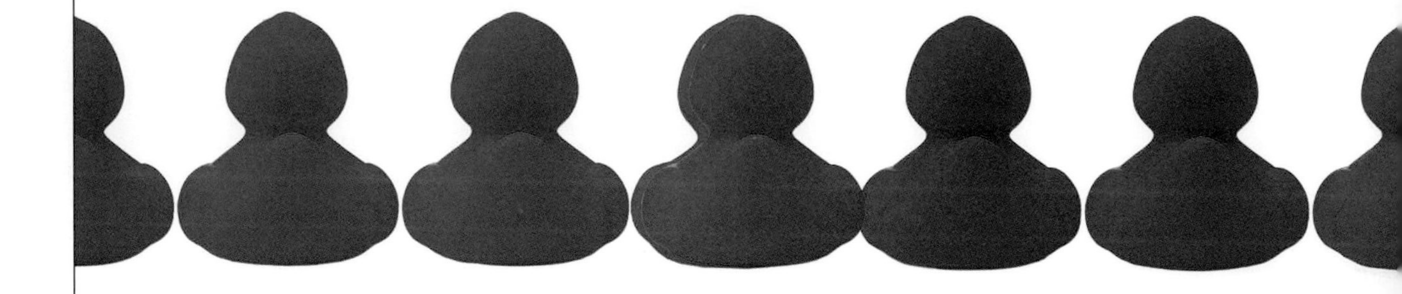

of authority figures. Opinions, attitudes, and behaviors move through a group in waves. Think of fashion trends, Internet memes, or political opinions. Even UFO sightings or outbreaks of seizures appear in clusters. Cultural norms can also change rapidly. In the 13 years between 2001 and 2014, for example, popular support for same-sex marriage rose from 35 percent of the American public to 52 percent, with 35 states making it legal.

People are suggestible. We adjust our thinking to group norms without even being aware of the change. Many classic experiments have shown this in action. A 1936 study asked a group of men to estimate how far a dot of light had moved. Starting with their own unique estimates, the men gradually changed their measurements to the group average—even when they were given the opportunity to express the estimates alone. The groupthink effect remained even as the group's members changed; new recruits adopted the preexisting norm.

The same effect appeared in experiments conducted by Solomon Asch in 1951. An experimental subject was told that he and others in his group were taking a perception test. Which line among three was the same length as the standard line, shown for comparison? Unknown to the subject, the other members of his group were planted by the experimenter. Consistently, they

SOLOMON ASCH TEST

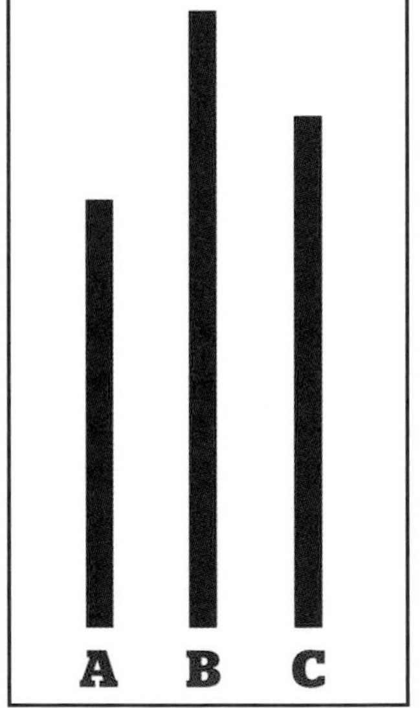

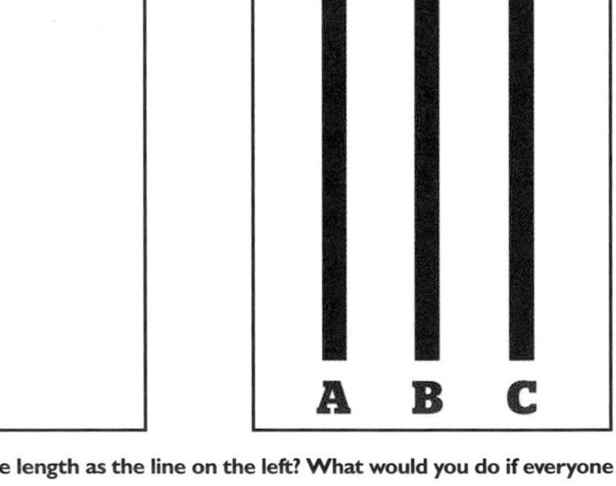

Which line is the same length as the line on the left? What would you do if everyone else in your group said "line A"?

identified an incorrect line as matching the standard. When the experimental subjects could answer the question on their own, they erred less than one percent of the time. When they were placed in a group, they shifted their judgment to the clearly incorrect norm one-third of the time.

Conformity starts very early; in playgroups, children willingly change their actions and preferences to match those of their friends. They dress, talk, and act to imitate the most popular kids. Even children who dislike broccoli will eat it if their friends do. As anyone knows who has ever attended high school, cliques are a defining social structure among adolescents, who self-organize into jocks, brains, rebels, and more.

As much as we might admire some nonconformists, few of us would want to live in a civilization without rules and social structure. Most people naturally respond to authority and to group consensus. We gain valuable information and safety as a member of a group. We also suffer painful consequences when rejected.

This tendency can make for a smooth-running society. It can also lead to horrifying abuses of power. The behavior of soldiers during the Holocaust is a notorious example. On two days in September 1941, for instance,

It's hard to be a nonconformist in a large group.

Nazi soldiers and Ukrainian police, under the direction of several commanders, systematically machine-gunned to death 33,771 Jews at the Babi Yar ravine in Ukraine. These troops were not a unique collection of monsters. They were ordinary Germans and Ukrainians who were "only following orders," to quote Adolf Eichmann.

Yet others have defied authority at the risk of their lives. Two years after Babi Yar, "White Rose" resistance member Sophie Scholl scattered leaflets from a balcony in Munich, urging defiance of Nazi rule: She was

"Your friends are God's way of apologizing for your relatives."

AUTHOR WAYNE DYER

guillotined by the Nazis days later. Legendary leaders such as Nelson Mandela and Mahatma Gandhi have brought about profound change through peaceful opposition.

Groupthink is a powerful force. It can express itself in organizations, such as the German army, or in spontaneous eruptions of violence, such as the riots following the acquittal of the Los Angeles police officers who, in their own eruption of group violence, beat construction worker Rodney King following a car chase. It doesn't always lead to abuse, of course, but it does create the mind-set that tolerates abuse. Experiments such as the Solomon Asch standard line test found that certain conditions increase conformity. Among them are group size (three or more people); group unanimity (everyone but you agrees); scrutiny (others are observing you); and insecurity (you feel incompetent, particularly in comparison to others in the group).

FOCUS

AUTHORITY & OBEDIENCE

Possibly the most famous psychological experiments of the 20th century, Stanley Milgram's studies of obedience are still dismaying in their implications. Milgram began them in 1961 in response to the "just following orders" explanations that arose at Adolf Eichmann's trial. In all, he conducted 20 tests with slightly different conditions, but the basic setup was as follows: Milgram recruited participants to his lab at Yale University, telling them that they were studying the effects of punishment on learning. Each participant was assigned to be the "teacher," while another subject, actually a confederate of the experimenter, was the "learner." In an adjoining room, the learner was wired with electrodes, while the teacher was told that every time the learner made a mistake in a word test, the teacher should shock him, increasing the level with each error. The shock levels visible to the teacher had labels ranging from 15 volts (slight) to 450 volts (severe and dangerous). In reality, the electrodes did nothing, but the learners had been told to behave as though they were in intensifying pain.

As the experiment proceeded, the learners deliberately made many errors. When the teachers increased the supposed shock level, the learners began to grunt, shout with pain, and at high levels, scream in agony and cry, "Get me out of here!" If the teacher refused to administer a shock, a white-coated authority figure ordered him to do so with "prods" such as "Please continue" or "You have no other choice but to continue."

Sixty-three percent of the teachers continued to shock the screaming learners right up to the highest level. Milgram was astonished by this result, but over the years, repeated tests with other subjects have confirmed his numbers. The experimenters found that certain conditions were most likely to prompt obedience:

- When the authority figure was at hand, wore a white coat, and assumed responsibility.
- When the experiments took place in official-looking surroundings.
- When the teacher could instruct someone else to press the switches.
- When the learner was in another room or at a distance.

On the other hand, in some circumstances teachers were much less likely to administer painful shocks:

- When the teacher had to physically force the learner's hand onto a shock plate.
- When other participants were seen refusing to obey. In those cases, obedience fell to only 10 percent.

»Cults

The human need for affiliation, the fear of rejection, and the tendency to yield to authority figures combine in their most extreme form in religious cults. Also known as highly restrictive religious organizations or high-intensity faith groups, cults represent a small, but not negligible, percentage of religious groups worldwide. In the United States, about two and a half million people belong to cults. Most such groups are small, containing fewer than 100 people, and attract little notice. Researchers Cynthia Matthews and Carmen Salazar, who have studied cults and interviewed their members, identified some criteria that distinguish cults from ordinary religious institutions. They include:

Gang graffiti in Honduras shows one way that groups establish dominance.

- **Patriarchy and gender roles.** Most (but not all) cults are male-dominated and hold to traditional patriarchal structures. Women are told to submit to male authority and are disciplined if they don't.
- **Sole decision-making.** Cult leaders make all the decisions for the group. A former cult member quoted by Matthews said, "Decisions were made by the leader—everything flowed by him. Guilt, shame, and shunning occurred if you hadn't accepted him as your leader or accepted his decisions."
- **Obedience to authority.** Cult members must submit without question to their leader's decisions.
- **Isolation.** Cult leaders use isolation from outside influences as a control tactic. In many cults, members are home-schooled and marry within the group. Outside influences are seen as evil.
- **Parental authority.** When entire families belong to a cult, fathers represent the power of the cult leadership. Typically, they use anger and punishment to enforce their children's

> **"When all think alike, then no one is thinking."**
>
> JOURNALIST WALTER LIPPMANN

Cultural moral systems provide a guideline for right and wrong.

obedience, while mothers employ guilt and shame.

• **Religiosity.** Cult members are often told that their path is the only way to salvation or enlightenment.

• **Abuse.** Psychological, emotional, physical, and sexual abuses in the name of the cult are among the most damaging characteristics of such organizations. All of the former cult members interviewed by Matthews and colleagues reported some form of abuse.

People who leave cults may grieve or feel angry for years. Many, however, draw on the support of others who leave with them and find healing by helping others to leave as well.

»Morals and Taboos

Some of our beliefs and behaviors as social animals are so widespread and ingrained that we view them almost as natural laws. Among them are moral rules. Philosophers continue to debate the origin of morals; in recent decades the subject has been reexamined by researchers ranging from evolutionary biologists to psychologists of various stripes.

Opinions continue to vary, but a persuasive school of thought holds that morality is a combination of intuitive mechanisms—formed by evolutionary pressures and to some extent hardwired into the brain—and cultural conditioning. Social psychologist Jonathan Haidt calls moral systems "interlocking sets of values, practices, institutions, and evolved psychological mechanisms that work together to suppress or regulate selfishness and make social life possible." Morality arises, he asserts, "from the coevolution of genes and cultural innovations."

Moral issues can assume greater or lesser importance, depending upon culture, but common themes emerge. They include:

- **Care/harm:** attachment systems and ability to feel others' pain, underlying the virtues of kindness, gentleness, and nurturance
- **Fairness/cheating:** evolutionary reciprocal altruism, underlying ideas of justice, rights, and autonomy
- **Loyalty/betrayal:** related to the ability to form social networks, groups, tribes, and underlying the virtues of patriotism and self-sacrifice for the group
- **Authority/subversion:** hierarchical social interactions, underlying the virtues of leadership and followership, including deference to legitimate authority and respect for traditions
- **Sanctity/degradation:** disgust and contamination, underlying the religious notions of striving to live in a more noble world; respect for our body that can be desecrated by immoral activities.

The dual nature of moral judgment—instinct plus cultural conditioning—is evident when we face moral dilemmas. Dealing with moral issues, people often

ASK YOURSELF

HOW MUCH DOES IT BOTHER YOU?

Every culture has its taboos—acts that seem disgusting or disrespectful, though usually harmless. Researchers have found that instinct, cultural conditioning, and social class shape how people view forbidden acts.

In one study, researchers posed the following scenarios to people in Brazil and in the United States and asked for their reactions. How much do these situations bother you—if at all?

» **A woman is cleaning out her closet, and she finds her old national flag. She doesn't want the flag anymore, so she cuts it up into pieces and uses the rags to clean her bathroom.**

» **A woman was dying, and on her deathbed she asked her son to promise that he would visit her grave every week. The son loved his mother very much, so he promised to do so. But after the mother died, the son didn't keep his promise, because he was very busy.**

» **A family's dog was killed by a car in front of their house. They had heard that dog meat was delicious, so they cut up the dog's body and cooked it and ate it for dinner.**

» **A brother and sister like to kiss each other on the mouth. When nobody is around, they find a secret hiding place and kiss each other on the mouth, passionately.**

» **A man goes to the supermarket once a week and buys a dead chicken. But before he cooks the chicken, he has sexual intercourse with it. Then he cooks it and eats it.**

Most people (in both the U.S. and Brazil) are not particularly disturbed by the "flag" or "deathbed" stories, but most are put off by the "dog," "kissing," and "chicken" scenarios, and felt the participants should be stopped or punished. Social class has a strong effect on reactions. People with higher socioeconomic status, such as college students, are the most permissive.

have a quick, intuitive reaction that doesn't live up to rational analysis.

A famous thought experiment illustrating this is the "trolley problem." In one of many versions, you are the conductor of a trolley speeding down the track. Just as the trolley's brakes fail, you see five workmen ahead of you who will certainly be killed unless you throw a switch that takes you onto an alternate track, on which stands a single workman, who would also be killed. Would you throw the switch?

Now imagine that you are standing on a bridge overlooking the trolley, which is out of control because the conductor has fainted. Next to you is a fat man. If you push him onto the track, he will be killed, but he will prevent the trolley from killing the five workmen. Would you push him over?

Our brain reacts to a friend's pain as if we're experiencing it ourselves.

Most people judging the first scenario would throw the switch; most in the second would not push the man off the bridge. Yet the second option has exactly the same results as the first: One man dies to save five others. People seem intuitively to make a distinction between intentional harm and harm as a side effect—between killing someone and letting someone die—regardless of outcome.

When it comes to moral dilemmas or taboo situations, experimenters have found that people tend to base their judgments on their first, gut reactions, and then to rationalize those judgments later. Such was the case with the trolley problem and the five scenarios in the sidebar on page 135, as well as with other questions involving sex, food (would you drink from a glass after a sterilized cockroach had been dipped in it?), or superstition (would you sign a nonbinding contract selling your soul to the devil for two dollars?). When questioned, many people are hard put to explain their aversion to these actions, while maintaining that there must be a reason. This behavior has been called "moral dumbfounding": We just know something's wrong, but we can't explain why.

Our need to connect with others shapes not only each individual life but also human culture as a whole, for good or for ill. "No man is an island," wrote John Donne. Psychologists agree. Recognizing the power of our social attachments can help us understand how strongly we are influenced by others—and how we can reject harmful influences and stand up for our own beliefs as well.

The "trolley problem" is a classic thought experiment that reveals moral judgment.

» CHAPTER FIVE «

RETHINKING INTELLIGENCE

As the sun was setting late one afternoon in 1881, Croatian engineer Nikola Tesla was walking through a park in Budapest, quoting a passage from *Faust* to his companion. "As I uttered these inspiring words the idea came like a flash of lightning and in an instant the truth was revealed," Tesla wrote later.

"I drew with a stick on the sand the diagrams shown six years later in my address before the American Institute of Electrical Engineers . . . The images I saw were wonderfully sharp and clear and had the solidity of metal and stone, so much that I told him: 'See my motor here; watch me reverse it.'" What Tesla had drawn in the sand was the induction motor, powered by alternating current, the foundation of today's worldwide electrical power.

Few people today would dispute that Tesla was a genius—but what does that really mean? A university dropout, Tesla had studied electrical engineering in school. Did his analytical skills and accumulated knowledge lead him to his answer? Was his poetry-fueled insight a moment of pure intuition, drawing on some deeper well than conscious thought? Did he possess far greater creativity than the average person—or was he simply more persistent, having pondered the issue of electrical motors for years?

All of these abilities and more may have contributed to Tesla's accomplishments, because intelligence today has a much broader definition than book smarts. According to contemporary researchers, being "smart" means being able to get where you want to go: It is the dynamic mix of ability and engagement in the pursuit of your goals. It can be a conscious process, an intuitive one, or a mix of both. Intelligence is not a solitary aptitude that exists only inside your head. Along with creativity, it's the way that your mind engages with the outside world to help you survive and thrive.

Engineer Nikola Tesla in his Colorado Springs laboratory

THE CHANGING MEANING OF INTELLIGENCE

The idea that intelligence is a single ability has long had its supporters and detractors. In the early 20th century, English statistician Charles Spearman proposed the idea that every person has a certain amount of underlying general intelligence, known as *g*. He noted that, although people differ in their specialized skills, a person who scores well in one area, such as vocabulary, will usually score above average in other skills as well. A certain level of intelligence seems to apply across the board. Other researchers have pointed out that *g* works mainly for analytical skills involving novel problems, such as those you might find in school, but not to other real-world situations.

For most people in modern society, intelligence comes down to numbers. IQ tests, state standardized tests, the SATs, and the like assign values to a range of academic skills as measured in a classroom or lab. These scores are gateways to advanced classes, colleges, and graduate schools. They are reasonably good predictors of academic success, but all too often, they become labels that children struggle to peel off. "She has an IQ of 110," a teacher might say, as if the score is an inborn trait, like brown hair. But a score on a standardized test is just that. It is simply one score on one test on one particular day, measuring a limited set of analytical skills. It makes more sense to say "she scored 110 on the test when she was ten," just as you might note that "she got a B on her algebra test last week."

Standardized tests typically measure achievement, aptitude, or both. Achievement reflects what you have learned, such as vocabulary or general cultural knowledge. Aptitude predicts how well you *will* learn, assessing such skills as problem-solving or spatial reasoning. The most commonly used adult intelligence test, the Wechsler Adult Intelligence Scale (WAIS), consists of four sections testing verbal comprehension, perceptual reasoning, working memory, and processing speed. Critics of standardized tests point out that these sorts of questions often have a built-in socioeconomic bias that disadvantages test-takers who aren't familiar with, say, the theater, or insurance. They note that people who tend to freeze up during tests—and we all know people who don't "test well"—will underperform. Self-perceptions affect test scores, too. Women who are told that women and men score equally well on a particular test will perform better than women who aren't given that assurance. Tests given in early childhood can't take into account a child's later experiences, events that can stimulate parts of the brain tied to cognitive ability.

Those who design these tests are well aware of the criticisms and attempt to create as level a playing field as possible. When used correctly, the tests do roughly predict achievement in school and in certain fields. However, no one test can capture the broader aspects of intelligence

THE FLYNN EFFECT

Are you smarter than your grandmother? She might not think so, but IQ scores tell a different story. Over the last century, scores in countries from China to Brazil have steadily increased, rising between 5 and 25 points per generation. It's a baffling phenomenon known as the Flynn Effect for James Flynn, the New Zealand researcher who first pointed it out.

Typical IQ tests, such as the Stanford-Binet or the Wechsler, standardize their scores across a population. A score of 100 represents the average, the midpoint on the bell-shaped curve of recorded scores. Testers have found that they must periodically reset the midpoint to reflect rising scores. By today's standards, the average score one hundred years ago would have been a 76—a number now considered a sign of intellectual disability.

This effect has been confirmed across a variety of tests, countries, and years. It is most significant with culturally dependent questions, such as those involving vocabulary, but scores for purely mathematical or "fluid" intelligence are also climbing steadily. The cause, or causes, remains a mystery. James Flynn believes it's a cultural effect. In his view, modern education, mass media, and a transition from farming to industrial economies have broadened our vocabularies and sharpened our analytical skills—essentially, molding our minds to the skills that are tested. However, these societal changes don't fully explain the increases in fluid skills. Improved nutrition may play a part, just as it has with increasing heights worldwide, but again, it doesn't seem to account for the size of the IQ boost. Smaller families, better health, or more simulating environments may also contribute to increased scores. Whatever the reason, the effect appears to be real.

that allow a person to thrive in his or her environment. A native of the Australian outback might score poorly on the SATs, for instance, but he would easily outdo an American college student at navigating a difficult desert environment.

» Multiple Intelligences

If intelligence is not one but several abilities, as most researchers now believe, just what are those abilities, how do we define them—and are they

IQ scores are rising 5 to 25 points per generation.

linked? American psychologist Howard Gardner identified eight separate intelligences: linguistic, logical-mathematical, musical, spatial, bodily-kinesthetic, intrapersonal (self), interpersonal

Logical-mathematical sense may be one kind of intelligence.

(others), and naturalistic. Although these categories make intuitive sense, Gardner's critics have pointed out that they don't all have scientific support.

An alternate theory, backed by testing, is Robert Sternberg's three-part, or triarchic, idea of intelligence. The components are:

- **Analytical intelligence:** the ability to complete academic, problem-solving tasks, such as those found in school or on

Every child has a unique set of learning skills.

1000000$
MARKET
20%
BUSINESS PLAN
WWW
2013
10,000,000€
SALE!

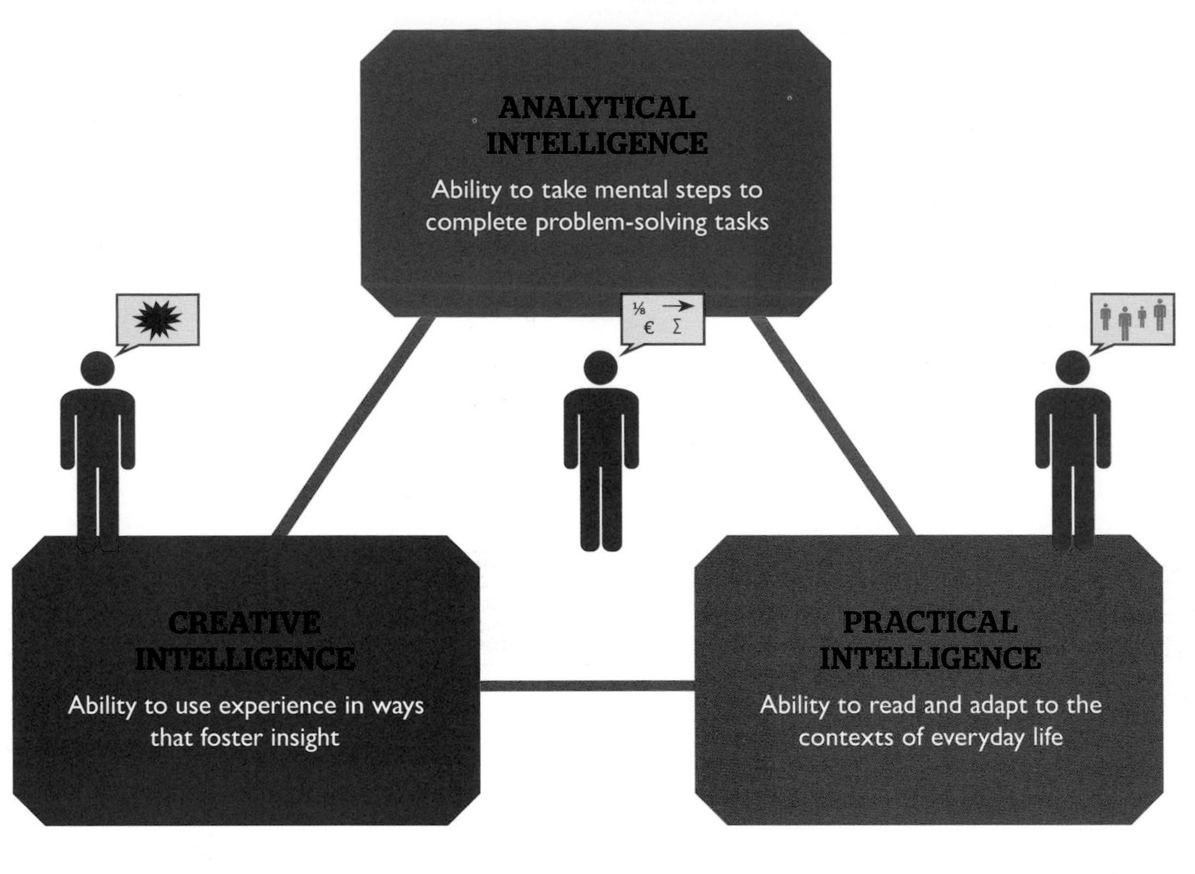

traditional intelligence tests—well-defined problems with a single correct answer. This ability, the closest to what we usually think of as intelligence, is used when we analyze, evaluate, or compare information. It's not all about book smarts, though; it applies to all sorts of everyday reasoning. Sternberg and colleagues, for instance, studied the difference between predicting and "postdicting"—imagining what something was like in the past. Given a firm peach, for example, what's your prediction for whether it will be firm or soft in a week? What's your guess for whether it was firm or soft a week ago? The researchers found that prediction comes faster, and easier, for people than postdiction. It's harder to analyze what may have happened in

"We know what we are but know not what we may be."

PLAYWRIGHT WILLIAM SHAKESPEARE, *HAMLET*

the past than to guess what will occur in the future.

• **Creative intelligence:** the ability to successfully deal with new and unusual situations and to come up with novel answers by drawing on existing knowledge and skills. People high in creative intelligence are apt to be innovators, good at discovering or inventing new ideas or products.

• **Practical intelligence:** the ability to adapt to everyday life by drawing on existing knowledge or skills. People with high practical intelligence are can-do types. They understand what needs to be done in a particular situation and then they tackle it. We might say they possess a lot of tacit knowledge, the kind of know-how that is picked up through experience but can't easily be explained. Riding a bicycle, cooking an omelet, or selling a car involves tacit knowledge.

According to Sternberg and colleagues, every human being possesses all three kinds of intelligence, but to different degrees. Some individuals have high skills across the board, but more typically a person will be stronger in some areas than in others.

All three abilities can be applied to a wide range of situations, from the schoolroom to the workplace to the home.

So, from this three-part perspective, can we simply say a person is intelligent? In a way, we can. We can identify intelligent people through their actions and certain key qualities. An intelligent person:

• **Can name and achieve his or her goals in life.** This recognizes that intelligence means different things to each person. A woman wanting to become a police officer will take an alternate path from the man who's aiming to become a distinguished novelist. Intelligence is also tempered by a person's

TRY IT

THINKING OUTSIDE THE BOX

One sign of creative intelligence is the ability to solve novel problems. Take a gander at the following questions, adapted from tests created by psychologists Robert Sternberg and Joyce Gastel. How quickly can you answer each one? Which ones are harder?

1. Chalk is used for writing. Therefore, ink is to paper as chalk is to:

a) word b) blackboard c) eraser d) classroom

2. The hand is the organ of hearing. Therefore, eye is to blinding as hand is to:

a) smooth b) touching c) deafening d) worn

3. Some people are heavier than others. Which answer completes the following series?

Skinny, slim, average, plump . . . a) hungry b) thin c) fat d) athletic

4. Furniture is eaten at the end of the meal. Which answer completes the following series?

Appetizer, soup, salad, main course . . . a) table b) bread c) menu d) entrée

ANSWERS

1 b, 2 c, 3 c, 4 a. Questions 2 and 4 represent novel problems. Solving them enlists your creative intelligence.

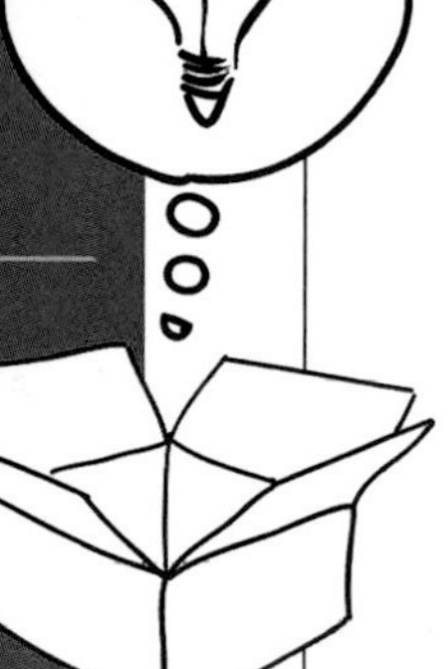

social context. It's a lot harder to become a distinguished novelist if you're from a poor household with little educational opportunity. However, the intelligent person will have the skills to reach the goal, or at least a plan for acquiring those skills.

• **Capitalizes on strengths and corrects or compensates for weaknesses.** If you can identify these characteristics and work within the pattern of your abilities, it's a sign of intelligence. How much attention you need to give your weaknesses depends upon how damaging they really are. Dr. Robert Biswas-Diener compares weaknesses to holes in the hull of a boat. If your life is the boat, sailing toward your desired destination, your weaknesses are gaps in its hull. Minor gaps will let in a little water, but not enough to truly slow you down. If you spend too much time fussing over them, you won't get to your destination. A bigger weakness, a larger hole, will require some attention and energy. You'll need to fix it to keep going. Sometimes, the weakness is so great that it threatens to sink the boat and end your journey. Then, you need to pull the boat out of the water and attend to it before you continue—or in other words, ignore your strengths and your goal for a little while until your weakness is dealt with.

Dr. Robert Biswas-Diener compares life to a boat on its way to a destination.

The first option that comes to mind is usually the best.

Some professions recognize different kinds of strengths and structure their jobs accordingly. In the United Kingdom, for instance, a lawyer may be a solicitor or a barrister. Solicitors are attorneys who deal with clients and work in the office, analyzing and writing up cases. Barristers are tapped for their speaking skills and appear in court, making a persuasive argument.

• **Adapts to, shapes, and chooses environments.** An intelligent person will change with a changing scene. A president, for instance, exhibits this kind of intelligence when he can lead equally well in wartime or peacetime. Intelligent people may also shape their environment so that it suits them, or leave one situation for another one that uses their strengths more wisely. A corporate executive, comfortable with people but miserable in the office routine, might show intelligence by shifting her profession to teaching.

• **Uses all three kinds of intelligence—analytical, creative, and practical—to reach a goal.** We do better when we recognize our varying skills and employ all of them.

"The man of true genius never lives before his time."

SCIENTIST JOSEPH HENRY

CONSCIOUS AND UNCONSCIOUS THOUGHT

When the diagram of the induction motor popped, unbidden, into Nikola Tesla's mind, it was probably produced by an invisible thought process known as unconscious cognition. We tend to think of the workings of intelligence as conscious and deliberate, involving the step-by-step problem-solving that goes into reading a map or adding a string of numbers. And indeed, some of our thinking, our conscious cognition, takes this form. However, a vast amount of information processing takes place below the level of conscious awareness. This cognitive unconscious is a place of quick judgments and rapid perceptions that affect our thoughts and behavior.

One of the most important roles of the cognitive unconscious is that of censor. Our brains screen out the innumerable sensory details that would otherwise drive us mad. Think of driving a car: If you had to attend to every tiny movement you made, every other car, road sign, building, cloud in the sky, tree by the side of the road, noise from outside, and the squirming of your kids in the backseat, you couldn't make it down the street. In fact, longtime drivers often find that they have so successfully blocked out the passing scene that they've gone for blocks on autopilot without registering the environment at all. This filter is known as latent inhibition. It's the brain's ability to screen out information that's irrelevant to the task at hand. People who have faulty mental filters are distractible, scattered, or worse: Schizophrenics, for instance, are deficient in latent inhibition, which means it's hard for them to block out unnecessary signals from their surroundings. As a

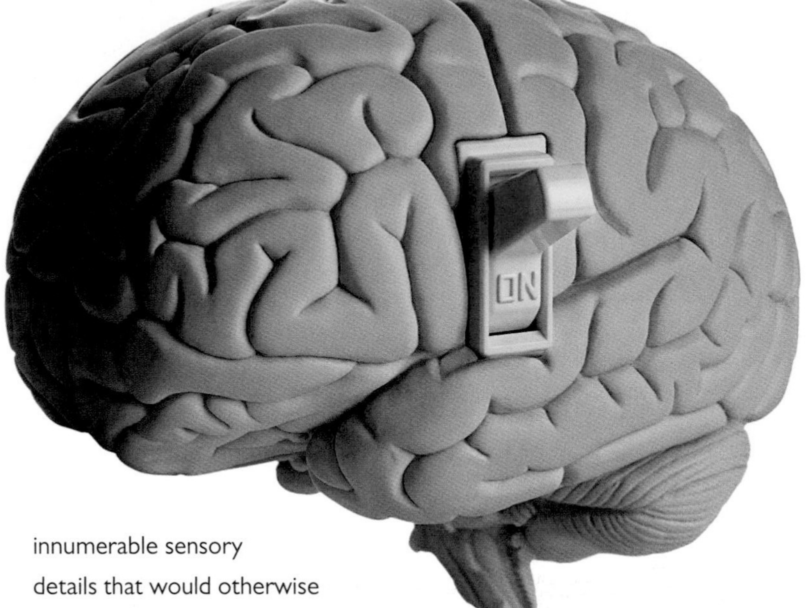

If our brains didn't block out distracting sensory information, we wouldn't be able to function.

"Wisdom and deep intelligence require an honest appreciation of mystery."

POET THOMAS MOORE

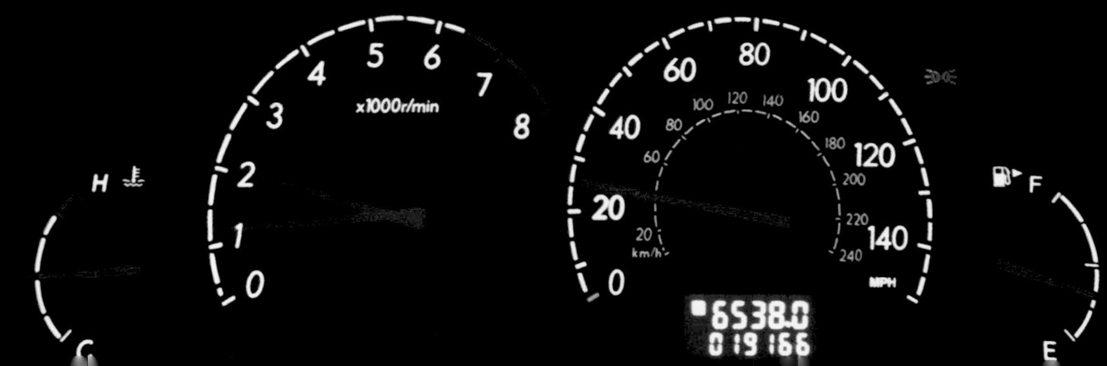

H
C
0 1 2 3 4 5 6 7 8
x1000r/min
0 20 40 60 80 100 120 140
MPH
20 60 80 100 120 140 160 180 200 220 240
km/h
6538.0
019166
F
E

result, they are bombarded by a blizzard of sensory noise that makes it tough to function.

The cognitive unconscious also serves as a learning center. Early science fiction used to depict futuristic homes in which adults could learn a foreign language or a wiring diagram simply by hearing tapes while they slept. Although this painless route to knowledge has, sadly, not yet been realized, scientists have discovered that the mind can process information without our conscious participation. As a result, we sometimes comprehend things quickly, almost instinctively, without thinking them through.

We may not be able to learn a language overnight, for instance, but we do grasp grammatical rules and patterns without consciously realizing it. In one experiment, participants were shown strings of letters arranged in patterns that reflected new and difficult grammatical rules. After a while, they were able to distinguish between "grammatical" and "ungrammatical" patterns in new letter strings, even though they couldn't consciously identify the rules that governed that grammar. Somehow their minds,

FOCUS

MORAL ALGEBRA

Benjamin Franklin, statesman, inventor, and writer, was a methodical thinker. In a 1772 letter, he described his decision-making process to English scientist Joseph Priestley:

"When these difficult Cases occur, they are difficult chiefly because while we have them under Consideration all the Reasons pro and con are not present to the Mind at the same time; but sometimes one Set present themselves, and at other times another, the first being out of Sight. Hence the various Purposes or Inclinations that alternately prevail, and the Uncertainty that perplexes us.

"To get over this, my Way is, to divide half a Sheet of Paper by a Line into two Columns, writing over the one Pro, and over the other Con. Then during three or four Days Consideration I put down under the different Heads short Hints of the different Motives that at different Times occur to me for or against the Measure. When I have thus got them all together in one View, I endeavour to estimate their respective Weights; and where I find two, one on each side, that seem equal, I strike them both out: If I find a Reason pro equal to some two Reasons con, I strike out the three. If I judge some two Reasons con equal to some three Reasons pro, I strike out the five; and thus proceeding I find at length where the Ballance lies; and if after a Day or two of farther Consideration nothing new that is of Importance occurs on either side, I come to a Determination accordingly.

"And tho' the Weight of Reasons cannot be taken with the Precision of Algebraic Quantities, yet when each is thus considered separately and comparatively, and the whole lies before me, I think I can judge better, and am less likely to take a rash Step; and in fact I have found great Advantage from this kind of Equation, in what may be called Moral or Prudential Algebra."

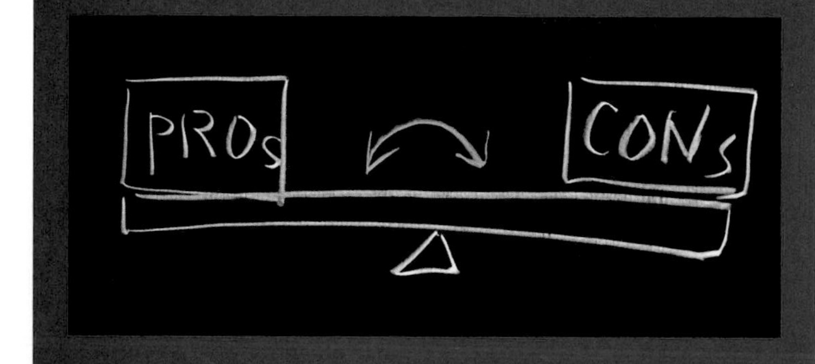

“Execution is the chariot of genius.”

POET WILLIAM BLAKE

through experience, had worked out the patterns and learned to recognize them.

Even under anesthesia, the brain can learn. In another experiment, researchers played audiotapes of word pairs to 25 anesthetized patients undergoing cardiac surgery. Afterward, none of the patients remembered the experience, but they did recall the word associations. Something in their brains had not only heard the tapes, but also learned from them.

»Mental Shortcuts

Unconscious cognition not only helps us learn, but also it helps us make rapid decisions through the mental shortcuts known as heuristics. Faced with choices, orderly people such as Benjamin Franklin (see sidebar, opposite) might write down pros and cons, but most of us rely upon rules of thumb to reach a conclusion. Certain answers just seem right to us.

Too much information can get in the way of decision-making by bogging us down in data. Heuristics, at their best, save us time and help us make more accurate choices. Researchers have identified various kinds of heuristics that allow us to cut corners and conserve energy. When we employ the recognition heuristic to choose between two options, for instance, we place a greater value on the option that's familiar to us. Studies have found that stock portfolios whose contents are chosen simply based on recognition value outperform the market. Recognition value can even predict athletic success. The degree to which tennis fans recognized amateur tennis players turned out to be a better predictor of the players' 2004 and 2006 Wimbledon matches

than their seeds or rankings.

What if both options are familiar? In that case, you might be better off going with the one that sprang to mind first. This is known as the fluency heuristic. For example, researchers showed experienced handball players video clips from a professional game and asked what they would have done in particular situations (pass the ball; take a shot). On average, the first option that came to their minds was better than those they thought of later, with more time to consider. The same effect has been seen with basketball players.

Explicit subliminal messages ("Drink Coca-Cola") do not work well.

Athletes use mental shortcuts to make quick decisions.

Athletes also use a common heuristic known as the one-clever-cue. Faced with multiple factors—let's say, a baseball player's stance, the wind direction, and the trajectory, speed, and spin of his line drive—a fielder simply maintains a constant optical angle between himself and the baseball. In other words, he eyeballs it.

Heuristics, as Ben Franklin would note, have their pros and cons. Just because an option springs to mind first, or is more familiar, doesn't mean that it gives us the most accurate information. Many things are familiar to us because they are sensational or frightening. People read about lottery winners and vastly overestimate the odds that they, too, will win the jackpot. They fear the statistically safe airplane more than the dangerous automobile.

Our thinking can be biased by sensational or frightening information.

Unconscious decision-making works best when we have to integrate large amounts of information quickly. Conscious thought is better when we need to follow strict rules and eliminate unlikely possibilities. When it comes to most complex decisions, we're better off using both skills. Studies have found that people who approach complex problems by first reasoning through them logically, then taking into account unconscious preferences, solve the problems far better than those who use either strategy on its own.

» Submerged Emotions

So much information processing occurs beneath the conscious

"By his genius distant lands converse and men sail unafraid upon the deep."

EPITAPH ON THE TOMB OF INVENTOR REGINALD FESSENDEN

Claims for subliminal advertising in the 1950s turned out to be exaggerated.

surface of our brains because we would be overwhelmed if we had to consciously attend to it all. The same is true for emotional processes. Although we're not aware of it, our emotions also respond to the world at a level below consciousness. They can drive our behavior and our decisions without our knowledge.

Many people have heard of an experiment performed by market researcher James Vicary in 1957. According to Vicary, during a screening of the movie *Picnic* he flashed the messages "Drink Coca-Cola" and "Hungry? Eat popcorn" in the film every five seconds for 1/3000 of a second—a level far below the limits of conscious perception. His claims that sales of Coke climbed 18.1 percent and those of popcorn rose an amazing 57.8 percent created a public sensation when they were reported. Unfortunately for popcorn makers everywhere, it turned out that Vicary had lied about his results. No one else could duplicate them. Subliminal advertising simply doesn't work on that scale.

Subliminal signals can, however, affect our emotions in more subtle ways. Positive or negative images, flashed too quickly to be consciously perceived, do

"I have also considered many scientific plans during my pushing you around in your pram!"

PHYSICIST ALBERT EINSTEIN, IN A LETTER TO SON HANS ALBERT EINSTEIN

change people's moods for better or worse. For example, two studies asked participants to rate their reactions to neutral images (Chinese symbols they had never seen before). Before each symbol was shown, a happy or an angry face appeared briefly, too quickly to be knowingly seen. Happy faces increased the participants' preferences for the symbols, while angry faces decreased preferences, even when the subjects were told that hidden pictures might affect their judgment.

Explicit subliminal messages ("Drink Coca-Cola") may not work well, but emotional cues can definitely affect consumption, particularly if the consumer is leaning that way anyway. Experimenters who flashed subliminal images of happy, neutral, or angry faces to thirsty participants found that the subjects didn't report any change of mood. Even so, the ones who viewed happy expressions drank 50 percent more of a fruit-flavored drink afterward and were willing to pay about twice as much for the drink.

It seems contradictory to say that we have feelings that we can't feel. It might be more accurate to say that we respond to all our feelings, but we're aware of only some of them. This is reasonable from an evolutionary viewpoint. Emotions are useful survival tools. They alert us to changes in our environment, for good or ill, and trigger an appropriate response. We fear and dislike things that threaten us. We like, and are attracted to, things that help us. But like other sensations, if we responded emotionally to every little thing we'd turn into a jittery mass of nerves. We need to

react to the environment without overreacting.

CREATIVE THINKING

One word can logically be paired with the following terms: "crab," "sauce," and "pine." What is it?

Now think about what you did while you considered this problem. Did you look at each word in turn and systematically work through the possibilities that matched each one? Or did you stare off into the distance as you free-associated the answer?

Researchers have found that we are more likely to experience flashes of insight—creative bursts of awareness—when we allow our minds to wander. We can certainly reach a solution through systematic reasoning, but we shouldn't suppress the fertile meanderings of our brains. They might effortlessly present us with the word "apple."

No one can stay focused and mindful every minute of the day. Think of the last time you were in a meeting. Did you pay attention to every word? It's not likely.

ASK YOURSELF

WHEN ARE YOU MOST CREATIVE?

"An intimate diary is interesting especially when it records the awakening of ideas," wrote author André Gide in his journal.

You can nourish your own newly awakened ideas by keeping a creative thoughts notebook, whether in paper form or on your smartphone or tablet. The next time an imaginative notion strikes you—an idea for a book, an art project, a new business, or whatever catches your fancy—jot it down as soon as you can before the idea flies out of your head in the business of everyday life. As you write, ask yourself what you were doing when the idea struck you, and note that down as well. Were you out for a walk? In the shower? Eating lunch? You might find that your most creative moments occur when your mind is idling.

KEEP AN INTIMATE DIARY

Boredom, distraction, and fatigue inevitably pull our minds away from the scene in front of us into a mental world of our own devising. This is not necessarily a bad thing. When our minds are wandering, they may be using a default mental network that processes information in novel ways.

Mind wandering can be a fundamental component of creativity. Creative ideas are original, innovative, and adaptive—useful in new ways. They don't come out of nowhere; Tesla couldn't have invented his induction motor without the years he put in studying electrical devices. But our default networks seem to discover and consolidate information that's already floating around in the brain. A study of professional writers and physicists found that more than 40 percent of their creative ideas occurred when they were involved in some non-work-related activity and/or were thinking about something unrelated to the topic. These "Aha!" moments were not necessarily more creative than those that occurred during deliberate cogitation, but they were more

Some of our best ideas come when we allow our minds to wander.

Daydreaming is important.

> **“Why is it that in most children education seems to destroy the creative urge?”**
>
> WRITER ALDOUS HUXLEY

likely to contribute to overcoming a mental impasse.

You may also be giving your emotions a boost when you allow your mind to wander, particularly if you're toiling through a boring task. People who take mental breaks find that the respite improves their outlook.

The best way to take advantage of a mental incubation process is not to do nothing, but instead to work away at some easy job. In one study, students were given a set of creative problems, asked to solve them, and then were assigned to either demanding tasks, undemanding tasks, a rest period, or no break at all. Afterward, the students who had been given the easy tasks performed better when they attempted the creative problems again. The period of mild mental activity had been an incubator for imaginative thought. The next time you're faced with a creative problem in need of a solution, try turning your hand to some undemanding activity, such as sorting out papers or talking a walk on the treadmill. See if the answers you need start to arise without being forced.

Nighttime dreaming resembles this kind of daydreaming, as explicit attention fades and the mind goes rambling down unexpected paths. Although scientists differ about the meaning and function of dreams (see chapter 9, page 259), many creative people look to their sleep for inspiration. Some even achieve the kind of eureka moment that Tesla experienced in the Budapest park.

Neuroscientist Otto Loewi, for example, recounted a now famous case of dream inspiration. In 1920, Loewi was studying the chemical transmission of nerve signals. One Saturday night

Take a break for creative activity to improve your work performance.

he dreamed of an experiment that could demonstrate this kind of transmission. "I woke, turned on the light and jotted down a few notes on a tiny slip of thin paper," Loewi wrote later. "Then I fell asleep again. It occurred to me at six o'clock in the morning that during the night I had written down something most important, but I was unable to decipher the scrawl." The next night, feeling desperate, he fell asleep and again dreamed of the new experiment. This time, he woke, remembered the procedure, and immediately repaired to his laboratory to try it out. The successful test, which identified the chemical transmitters acetylcholine and adrenaline, subsequently won Loewi the Nobel Prize.

The effects of Loewi's dream did not end there. A generation after Loewi, New Zealand physiologist John Eccles studied neurotransmission as well. His theory of how electrical signals traveled between cells also came to him in a dream. "On awakening I remembered the near tragic loss of Loewi's dream so I kept myself awake for an hour or so going over every aspect of the dream, and found it fitted all experimental evidence," Eccles wrote. He, too, won a Nobel Prize for his work. Even for those among us not in line for a Nobel Prize, keeping a dream journal and noting down our nighttime thoughts (clearly!) can lead to creative discoveries.

FOCUS

THE GUITAR CASE & THE PEACOCK'S TAIL

Who's the sexiest: the guitarist, the jock, or the average Joe?

French researchers tested this by sending an attractive young man onto the streets of a midsize city with either a guitar case, a sports bag, or nothing at all in his hands. He approached young women with a smile and said, "Hello. My name's Antoine. I just want to say that I think you're really pretty. I have to go to work this afternoon, and I was wondering if you would give me your phone number. I'll phone you later and we can have a drink together someplace."

Almost one-third of the young women gave their phone numbers to "Antoine" when he was carrying the guitar case, as opposed to 14 percent when he was holding nothing, and only 9 percent when he held a gym bag.

Is creativity sexy? Do women see it as a sign of evolutionary strength? Researchers from the time of Darwin have speculated that creativity is like the peacock's tail: an indicator of health in the competitive realm of sexual selection. Just as extravagant plumage shows that a male peacock has resources to spare, so creativity shows that a person has mental resources beyond the basics.

Creative people, particularly those in the arts, report more sexual partners than their less creative counterparts. However, we do tend to prefer people with similar interests to our own, so men and women who value technological creativity will choose that kind of ability in a partner. As for the jocks with their gym bags—they might want to swap them out for guitars.

»Creative People, Creative Processes

The stereotype of the bohemian,

flamboyant artist is not wrong. Creative people, in general, tend to be open to new experiences, flexible, bold, independent, and unconventional. This is particularly true for artists and less so for scientists, but even technologically creative people are more open and independent than the average. A sense of humor is also linked to creativity. Both humor and inventiveness involve playfulness, novelty, and the ability to make surprising connections between concepts.

Creativity is a preferred factor in sexual selection.

Studies show that humor itself is a creative process. Funny people are typically creative, although the reverse is not always true: Creative people can be humorless.

Whether you think of yourself as artistic or not, you can nurture your creative output by understanding how the creative process works. The first rule is not to wait for inspiration to strike, because inspiration comes after, not before, the creative idea. Once you have a creative idea, inspiration steps in to realize it. And inspiration itself can take three forms: transcendence, evocation, and approach motivation. Transcendence allows a person to see possibilities in her creative idea that go beyond the ordinary. Evocation is inspiration triggered by another person's creative idea or act. Approach motivation is the desire to act on the creative idea, to express it and make it real. When they're inspired, creative people start with an idea, possibly triggered by someone else, then see its potential and want to put it into action.

Though it seems counterintuitive, the creative process thrives on constraints. Given a blank piece of paper and told to do absolutely anything with it, most people feel lost and tend to repeat what they've done in the past. With constraints come a sense of direction and possibility. Knowledge is one such constraint. A creator needs to have acquired some learning and skills in her field—all the talent in the world won't make a Michelangelo out of an untrained sculptor. She will also be constrained simply by the limits of her brain, since even the brightest of us can only process so much information. She will be constrained as well by the need for variability; a challenging task will stimulate more original thinking than a repetitive one. And she will be constrained by the limits of her ability. Some of us are simply more talented than others.

What about mood? Does a scowling, Ludwig van Beethoven–like intensity promote creativity, or should we aim for an upbeat, Benjamin Franklin–style positivity? Both, it turns out. Negative moods

FOCUS

DOES BRAINSTORMING WORK?

The brainstorming session is a workplace staple. A problem needs to be solved, a project begun, a campaign planned: Gather the team and brainstorm! Brainstorming as a business technique was popularized in advertising executive Alex Osborn's 1948 book, *Your Creative Power.* Key to the method's success, he wrote, was a positive, all-ideas-accepted approach. "Creativity is so delicate a flower that praise tends to make it bloom while discouragement often nips it in the bud." The never-say-no brainstorming style spread throughout businesses large and small in the decades that followed.

However, as author Jonah Lehrer notes, "there is a problem with brainstorming. It doesn't work." Study after study has shown that groups, on average, come up with fewer ideas than individuals working on their own. Furthermore, ideas generated in an all-positive groupthink experience tend to converge on one solution, rather than diverging creatively. The classic brainstorming session actually inhibits creative thought.

Fear not, fans of collaborative thought. Creative thinking can be promoted by placing people with different perspectives close to each other and making sure they run into each other frequently. Meetings should abandon the "no criticism" rule and allow for healthy debate. And companies should encourage people to brainstorm their own ideas, on their own, and then later pool the results with others.

Creativity needs a sense
of direction and possibility.

seem to promote concentration and precise execution, particularly when the task itself is framed as a serious one. Positive moods boost creativity when the task is viewed as fun and enjoyable.

Beethoven's rages and despairs fit the popular image of the artist as madman. There is some truth to this stereotype. Creativity is correlated with trauma and mood disorders in many artists, but it may also be the life jacket that helps those artists transcend their troubles. Intellectually gifted people typically come from nurturing homes, but highly creative people often have pain in their pasts. They are more likely to be the children of troubled marriages, to have lost one or both parents in childhood, or to have endured other distressing experiences. Their inherent talents may have allowed them to rise to their life challenges and become productive.

The lack of latent inhibition that can be so disabling for schizophrenics may also have its positive side in some people.

Vincent van Gogh's mental illness fueled his art, such as his masterwork "The Starry Night," painted while he was in an asylum.

The classic brainstorming session actually inhibits creative thought.

Those with attention deficit/hyperactivity disorder (ADHD) will typically perform worse than other people on standardized tests, but they tend to score higher on laboratory measures of creativity and may achieve more than others in creative pursuits. Mood disorders such as depression or bipolar disorder, though crippling in many ways, are eight to ten times more prevalent in writers and artists than in the general population. German composer Robert Schumann, who suffered from bipolar disorder, cycled between manic periods and depressive episodes so severe that he suffered from hallucinations and attempted suicide by jumping off a bridge. In his manic periods, he completed about four times as many works as he did during depressive episodes. However, in the judgment of most musicians, the quality of his work did not improve with the quantity; the music he wrote during depressive moments is considered to be as good as the music of his manic times. American poet Theodore Roethke, another bipolar sufferer, wrote some of his strongest works in the wake of his breakdowns. "In a dark time, the eye begins to see," one poem begins.

Ludwig van Beethoven

"Imagination is more important than knowledge."

PHYSICIST ALBERT EINSTEIN

Few of us possess either the genius or the madness of a Schumann or Roethke, but all of us have creative and intellectual abilities that can be fostered if we're aware of them. Understanding your own forms of intelligence—your own combination of analytical, practical, and creative abilities—will help you find the best path toward reaching your goals. Tapping into your mind's hidden resources and its ability to make connections can also help you find answers you didn't know you knew. When we allow our minds to step off the analytical track and tap into deeper, unseen associations, we can each bring out our own genius.

Reaching into your mind's hidden resources can free your untapped abilities.

» CHAPTER SIX «

WHAT DRIVES YOU?

On January 14, 2015, two American rock climbers pulled themselves over a rocky ledge and stood triumphantly at the top of El Capitan, Yosemite National Park's iconic rock formation. They had just become the first people to make a free ascent of the monolith's Dawn Wall, 3,000 feet of sheer granite implacability. Using solely the strength of their arms and legs to ascend, with ropes to aid them only if they fell, Tommy Caldwell and Kevin Jorgeson pulled themselves from one hair-thin crack to another for 19 days. Each section of the wall, or pitch, presented its own challenges. On one difficult pitch, Jorgeson fell, climbed, and fell again for ten days in a row. The climbers' fingers bled and their muscles burned. But the two men persisted, and a climb seven years in the planning was finally accomplished on a sunny afternoon.

What was it that drove these climbers to attempt such a difficult feat? Why did they succeed? Personality psychologists might say that they scored high in the traits of extroversion, emotional stability, and openness. Those who study motivation would note that they were motivated to approach new experiences and actively reach toward their goals. And people who research self-control would agree that the climbers certainly possessed sky-high levels of persistence, or grit.

Few of us will have such a sharply defined moment in the sun. But all of us define ourselves every day by what we do, by the ways in which we interact with other people and with our environment. Our characters are shaped by our drives, our motivations, and our willingness to persist toward our goals. These personality traits, motives, and drives combine for a unique identity.

PERSONALITY

Thousands upon thousands of adjectives describe personality traits. Gregarious, kind, patient. Rude, lazy, aloof.

Thousands of adjectives are linked to personality traits.

Psychologists in the 20th century, attempting to organize scattered ideas about personality into something testable, began to see that certain adjectives correlated with others. Talkative people were more likely to be described as assertive. Moody individuals were often insecure. Researchers now recognize that these adjectives tend to cluster into five broad classes of personality traits, often called the Big Five: openness, conscientiousness, extroversion, agreeableness, and neuroticism (which can be remembered with the handy acronym OCEAN).

Every personality contains a mixture of all of these factors. Each trait can be measured on a scale from very low to very high. People very high in agreeableness might be described as sympathetic and kind; at the low end, harsh or irritable. The scales don't necessarily correlate: a disagreeable person can be highly conscientious, or a shy one open to new experience.

Personality traits can predict health outcomes.

When applied to large groups of people, the Big Five factors are linked to certain kinds of behavior and certain life outcomes. But it's important to know what they can't do. They can't predict individual responses in any given situation or describe every subtle variation in human personality. They represent broad categories, not individual life histories.

Across a population, personality factors have consistently been able to predict health outcomes. Not surprisingly, high conscientiousness is strongly linked to better health and greater longevity. People at the high end of that scale are more likely to follow healthy diets and use seat belts and less likely to smoke or binge drink. Highly extroverted people have good social networks, which are positive for health, but they are more likely to smoke, drink excessively, and have risky sex. Highly agreeable people are also likely to drink more than average, but those at the low, hostile end of the scale are more prone to cardiovascular disease. Anxious folks with high neuroticism scores experience generally worse health than others, ranging from eating disorders and substance abuse to tinnitus.

» Openness

The trait of openness reflects the tendency to seek out new experiences and knowledge. People who score high in openness are receptive to new ideas; they are more likely than others to go to

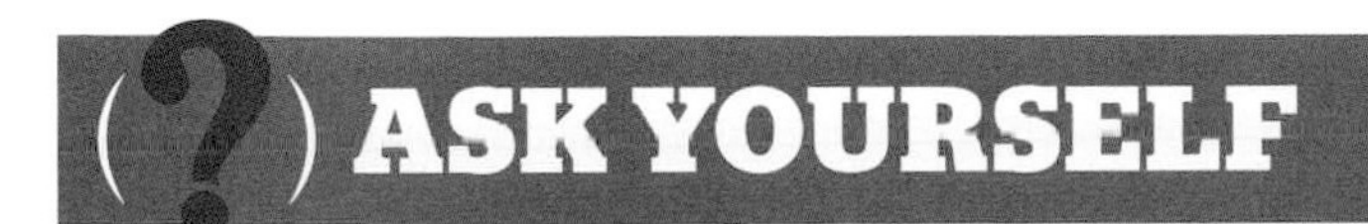

Where do you fall on the scales of the Big Five? Choose a number for each statement to indicate the extent to which you agree or disagree with it. The resulting score shows where you fall, from low to high on a scale of 1 to 7.

I SEE MYSELF AS:

1. _____ Extroverted, enthusiastic.
(1 = Disagree strongly 2 = Disagree moderately 3 = Disagree a little 4 = Neither agree nor disagree 5 = Agree a little 6 = Agree moderately 7 = Agree strongly)

2. _____ Critical, quarrelsome.
(1 = Agree strongly 2 = Agree moderately 3 = Agree a little 4 = Neither agree nor disagree 5 = Disagree a little 6 = Disagree moderately 7 = Disagree strongly)

3. _____ Dependable, self-disciplined.
(1 = Disagree strongly 2 = Disagree moderately 3 = Disagree a little 4 = Neither agree nor disagree 5 = Agree a little 6 = Agree moderately 7 = Agree strongly)

4. _____ Anxious, easily upset.
(1 = Agree strongly 2 = Agree moderately 3 = Agree a little 4 = Neither agree nor disagree 5 = Disagree a little 6 = Disagree moderately 7 = Disagree strongly)

5. _____ Open to new experiences, complex.
(1 = Disagree strongly 2 = Disagree moderately 3 = Disagree a little 4 = Neither agree nor disagree 5 = Agree a little 6 = Agree moderately 7 = Agree strongly)

6. _____ Reserved, quiet.
(1 = Agree strongly 2 = Agree moderately 3 = Agree a little 4 = Neither agree nor disagree 5 = Disagree a little 6 = Disagree moderately 7 = Disagree strongly)

7. _____ Sympathetic, warm.
(1 = Disagree strongly 2 = Disagree moderately 3 = Disagree a little 4 = Neither agree nor disagree 5 = Agree a little 6 = Agree moderately 7 = Agree strongly)

8. _____ Disorganized, careless.
(1 = Agree strongly 2 = Agree moderately 3 = Agree a little 4 = Neither agree nor disagree 5 = Disagree a little 6 = Disagree moderately 7 = Disagree strongly)

9. _____ Calm, emotionally stable.
(1 = Disagree strongly 2 = Disagree moderately 3 = Disagree a little 4 = Neither agree nor disagree 5 = Agree a little 6 = Agree moderately 7 = Agree strongly)

10. _____ Conventional, uncreative.
(1 = Agree strongly 2 = Agree moderately 3 = Agree a little 4 = Neither agree nor disagree 5 = Disagree a little 6 = Disagree moderately 7 = Disagree strongly)

RESULTS

Extroversion: add 1 + 6 and divide by 2
Agreeableness: add 2 + 7 and divide by 2
Conscientiousness: add 3 + 8 and divide by 2
Emotional stability (low neuroticism): add 4 + 9 and divide by 2
Openness: add 5 + 10 and divide by 2

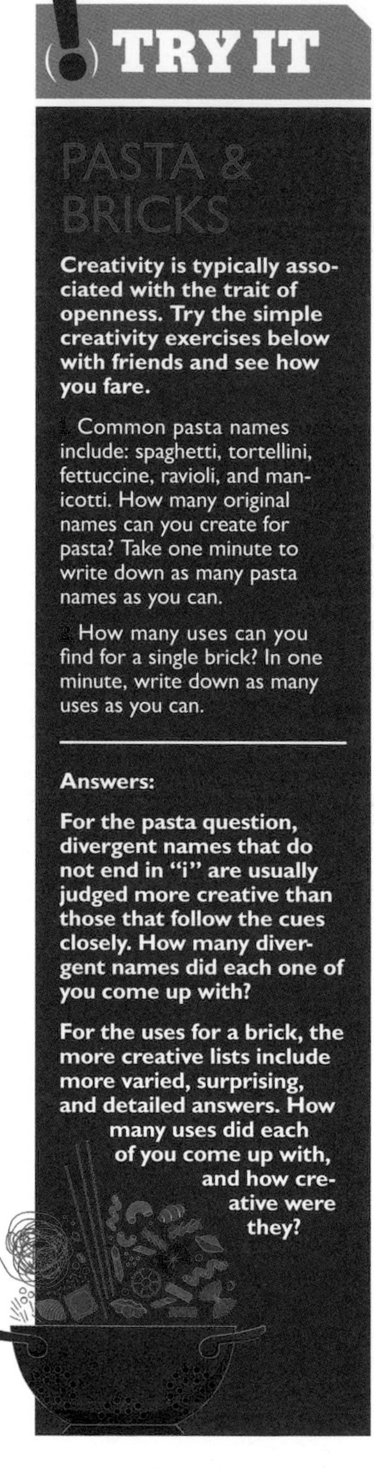

TRY IT

PASTA & BRICKS

Creativity is typically associated with the trait of openness. Try the simple creativity exercises below with friends and see how you fare.

1. Common pasta names include: spaghetti, tortellini, fettuccine, ravioli, and manicotti. How many original names can you create for pasta? Take one minute to write down as many pasta names as you can.

2. How many uses can you find for a single brick? In one minute, write down as many uses as you can.

Answers:

For the pasta question, divergent names that do not end in "i" are usually judged more creative than those that follow the cues closely. How many divergent names did each one of you come up with?

For the uses for a brick, the more creative lists include more varied, surprising, and detailed answers. How many uses did each of you come up with, and how creative were they?

art galleries or concerts, to learn something simply for the sake of learning, to try out new foods, or rearrange their living space just for the heck of it. To some extent this factor is correlated with intelligence and education, so some psychologists believe it should be broken into two categories: intellect and openness to experience. However, openness more fundamentally measures the breadth and originality of a person's mental life.

Outcomes predicted by high scores in openness include more years of education completed, more success in artistic jobs, more mobility from state to state, and even a greater likelihood of having tattoos. Highly open people also have a higher risk for bipolar disorder. They have more vivid dreams than others, and are more likely to remember their dreams.

Low scorers are more likely to be politically conservative. People high in openness are more willing to make changes, although they also suffer more stress because of those changes.

» Conscientiousness

As a personality factor, conscientiousness is just what it sounds like. People high in conscientiousness are industrious, organized, and reliable. They have excellent impulse control. Conscientious people are more likely than others to study for tests, make their beds, and plan ahead. A high score on the conscientiousness scale often predicts good grades at school, success at work, better health, and even longer life. However, highly conscientious people may also be inflexible and obsessive.

> **We owe about 50 percent of our personality to genetics.**

Those who score at the low end of the conscientiousness scale are typically impulsive, messy, and unwilling to think ahead. They often do poorly in school and at work. Low scores are connected to bad health habits such as smoking, substance abuse, and poor diet and exercise habits. Low-C folks are more likely to indulge in risky sexual behavior and to pursue multiple

People high on the openness scale are more likely to have tattoos.

romantic partners. Gamblers and criminals, unsurprisingly, tend to have particularly low scores in this area.

Although it may be harmful at the extreme high end, in general high conscientiousness makes for a smoother, healthier, more successful life.

You can't always judge personality from appearance.

»Extroversion

Carl Jung introduced the public to the terms extrovert and introvert, and the words are still commonly used to describe opposite personality types. Today, most psychologists recognize one scale for extroversion, encompassing the silent loner at one end and the party animal at the other. People who score high in extroversion enjoy social attention and experience strong positive emotions. (In one experiment, people high and low in extroversion were shown pictures of puppies while being scanned in an MRI. High-scoring people showed more brain activity than low scorers.) They are typically active, talkative, more likely than others to chat up romantic partners, and more inclined to take on leadership roles at work. Those low in extroversion are usually quiet, content with isolation, and more subdued emotionally.

Though sociable, highly extroverted people are not necessarily pleasant. Extroverts can also be aggressive and disagreeable. They may be more inclined to seek thrills at the expense of personal safety or stability. One study shows that extroverts are more likely to get into car

Extroverts enjoy social attention and are typically active and talkative, on occasion to the extreme, as seen here.

accidents while driving fast and listening to music.

Introversion has its strengths: Introverts may be more likely to listen to others, to work well independently, and to come up with thoughtful solutions. The famously extroverted Bill Clinton made it to the presidency of the United States, but so did the famously introverted Abraham Lincoln. Introverts can achieve as much as extroverts by leveraging their independence, thoughtfulness, and calm, steady natures. Albert Einstein, Isaac Newton, and T. S. Eliot were introverted. So are Bill Gates and J. K. Rowling.

Creative introverts can make productive use of their ability to work in quiet isolation. Entrepreneurial souls often work

PERSONALITY CONTINUUM SCALE

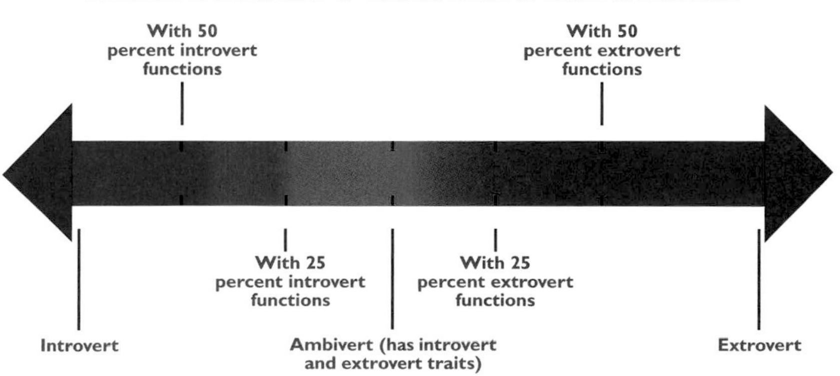

better independently, striking out in new directions without being influenced by a crowd. And introverts don't necessarily avoid public notice when they have something to contribute. Meryl Streep and Harrison Ford are often cited as introverts who are able to keep their screen personas separate from a quiet personal life.

» Agreeableness

Like conscientiousness, agreeableness is just what it sounds like. Those who score high in agreeableness are cooperative, empathetic, trusting, and modest. They avoid conflict and hostility. Highly agreeable people are sensitive to other people's feelings. They tend to speak well of others, to lend items to friends who need them, and to console others who are upset. They are more likely to donate to charities. People who do volunteer work typically combine high agreeableness and high extroversion. Highly agreeable people also do well working in groups and make successful politicians.

Highly agreeable people are friendly and trusting.

People who score low on the agreeableness scale can range from being indifferent to being actively aggressive and antagonistic. They are more likely to seek out conflict. Daily life in a household with a low-A person might include a lot of slamming doors, raised voices, and clenched fists. Low scores on the agreeableness scale are

FOCUS

FUNNY PEOPLE

Few performers are more exposed than stand-up comics. Not only do they have to create their own material, they must perform it in front of live audiences who provide instant feedback in the form of laughter or catcalls. In 2009, two psychologists published a study of professional comedians that measured them on the five-factor personality scale—with surprising results.

Predictably, comedians scored very high on the openness scale. In this, they are like actors and writers: creative and willing to try out new ideas. The stand-ups came in low on conscientiousness. Again, this wasn't surprising. Other studies have shown that low scores in this category are associated with negative or hostile humor styles. Unlike actors, they are low in agreeableness, a score perhaps linked as well to hostile or disparaging humor. But perhaps most surprisingly, comedians were about the same as the general population in neuroticism, but lower than others in extroversion. The typical stand-up comic is a stable, introverted soul.

"Human beings can alter their lives by altering their attitudes of mind"

PSYCHOLOGIST WILLIAM JAMES

linked to cardiovascular disease later in life.

» Neuroticism

The Big Five factor of neuroticism is sometimes called emotional stability. Scoring at the low end on the neuroticism scale indicates that you are typically on an even emotional keel, not inclined to worry or become too stressed by life events. High scorers are worriers: Moody and anxious, they respond strongly to negative events and constantly doubt their own actions and abilities. People high in neuroticism have more unstable relationships and job histories than others. (Note that scoring high in neuroticism is not the same as being neurotic. To psychologists, "neurosis" is an old-fashioned term for a mental disorder such as obsessive-compulsive disorder.)

> **Self-control can feel like a muscle; it's tired after exercise.**

High scores in neuroticism are linked to phobias, risky behavior, fatigue, and poor health. More inclined to worry about their health, people high on this scale are paradoxically less likely to protect it, although they are frequent visitors to the doctor. Stressful life experiences, such as the death of a loved one, are more likely to lead to emotional trauma. High scores are a significant risk factor for depression.

Are there any upsides to a high score on this factor? Studies find that people with moderately high neuroticism may be more realistic when assessing problems and potential dangers than

People high in neuroticism tend to be moody and anxious.

others. A sunny outlook is not always the sanest one.

Some scientists believe that people high in neuroticism have easily activated limbic systems (brain structures under the cerebrum that are associated with emotion). Levels of neurotransmitters, the chemicals that ferry information across brain cells, also seem to vary according to personality type in general. The enzyme monoamine oxidase (MAO), which breaks down neurotransmitters, has been shown to be lower than average in sensation-seeking personalities. It may be that these wild partiers and fast drivers have neurotransmitter levels that are simply too high for safety.

Knowing where you fit in to the big spectrum of personality types can help you with both acceptance and change. If you realize you are naturally introverted, for instance, you might stop beating yourself up about avoiding big parties and start appreciating the strengths of an introvert. If you're unhappy with your not-so-social life, you can also reflect that self-knowledge is the first step toward improvement. Nothing in your behavior is written in stone. You can acknowledge your natural tendencies and consciously work on redirecting them, a little at a time.

Easily stressed people may have overactive limbic systems.

» Inheriting Personality

Where does personality come from? Most people point to their upbringing: parents who encouraged or discouraged them, family dynamics that shaped their character. However, studies of personality point in a different direction—to genetics. Research involving twins bears this out.

Identical twins are rare, representing about 3 in 1,000 births. Identical twins separated at birth and raised in separate households are vanishingly rare, but they are much sought after by scientists attempting to untangle the influences of nature and nurture.

Studies of identical twins show that personality is genetically based.

Birth order has little effect on adult personality.

Consider a famous pair: the Jim twins. Adopted into different families at birth in 1939, the men (both named Jim) reconnected in 1979. They were physically similar, both 6 feet tall and weighing 180 pounds. Their personalities and life stories were also remarkably alike. Both had had dogs named Toy when they were growing up. Both had been married twice, first to a woman named Linda and second to one named Betty. Each smoked Salem cigarettes and drank Miller Lite. They had both worked as part-time sheriffs, enjoyed home carpentry, bit their fingernails, and left love notes for their wives around the house. Each had a son, one named James Alan and the other named James Allan. Standardized tests confirmed anecdotal observations; the two men had very similar personalities.

The Jims were only one example, and could have been a fluke, but studies of the population in general confirm that personality and intelligence are highly heritable. Personality seems so intangible—and is expressed through so many little behaviors and habits—that it's hard to believe that it is carried within our genes, but about 50 percent of variation in personality can in fact be attributed to genetics. Not only are the personalities of identical twins raised in separate households highly correlated, but their counterparts, adopted children raised together, show essentially no personality correlation with their siblings or adoptive parents.

Yet if 50 percent of personality is heritable, that still means that 50 percent comes from the environment. The logical place to look is at the shared setting of the household: Most people would agree that their personalities are shaped by their upbringing. Surprisingly, this is not borne out by research. Shared environments seem to have little impact on basic personality traits, as can be seen by the lack of similarity among adopted siblings. Even birth order has little effect on adult personality. The dynamic of the bossy, high-achieving firstborn and the irresponsible, black-sheep youngest can play out according to type when siblings are interacting within their family environment. However, it doesn't define their personalities or achievements outside the home. Firstborn children, for instance, fare no better in school than later-borns.

So what does account for the 50 percent of personality that isn't inherited? We don't know. The non-shared environment, such as the peer group, may have a strong influence. Perhaps a host of small factors come to bear. Or, quite possibly, we just don't know enough yet about how personality forms.

»Motives

Personality traits guide our interactions with the world, but there are other ways to understand why we do what we do. All of us have needs and desires. From these spring motives, our

FOCUS

DUBIOUS TESTS

Psychologists, human resources managers, prosecutors, and a host of other professionals use psychological tests to determine everything from a person's suitability for a particular job to criminal insanity. However, some of the most popular of these tests, including the Rorschach inkblot test and the Myers-Briggs Type Indicator, may be little more than intriguing ways to pass the time.

Devised by Swiss psychologist Hermann Rorschach, the inkblot assessment is a classic projective test. How a person interprets an ambiguous image is believed to reveal that person's personality and hidden emotions. It makes intuitive sense to conclude that the person who sees a dead parent in an image has different psychological issues from the one who sees a flower. In reality, though, the Rorschach test simply isn't scientific. Highly subjective scoring and no real statistical reliability or validation make it, at best, suggestive, but not predictive across a broad range of people.

Even more widely used, the Myers-Briggs Type Indicator (MBTI) also comes under fire from researchers. Its inventors, Katherine Briggs and her daughter Isabel Briggs Myers, had no scientific training, but based the test on their interpretations of the personality theories of Carl Jung. The MBTI uses yes-no questions to divide people into 16 combinations of introversion/extroversion, intuition/sensing, thinking/feeling, and judging/perceiving. Critics point out that the yes-no format and the arbitrary categories create a false dichotomy. A person will be rated as introverted or extroverted, a judger or a perceiver, but can't fall somewhere on the scale between the two labels. Moreover, the Myers-Briggs is inconsistent. Up to 50 percent of people who take the test twice come up with different personality types on the second try. Nor does the test predict success on the job particularly well, despite its use in workplaces around the country. It's popular with career counselors, but if you take the test, take it with a large grain of salt.

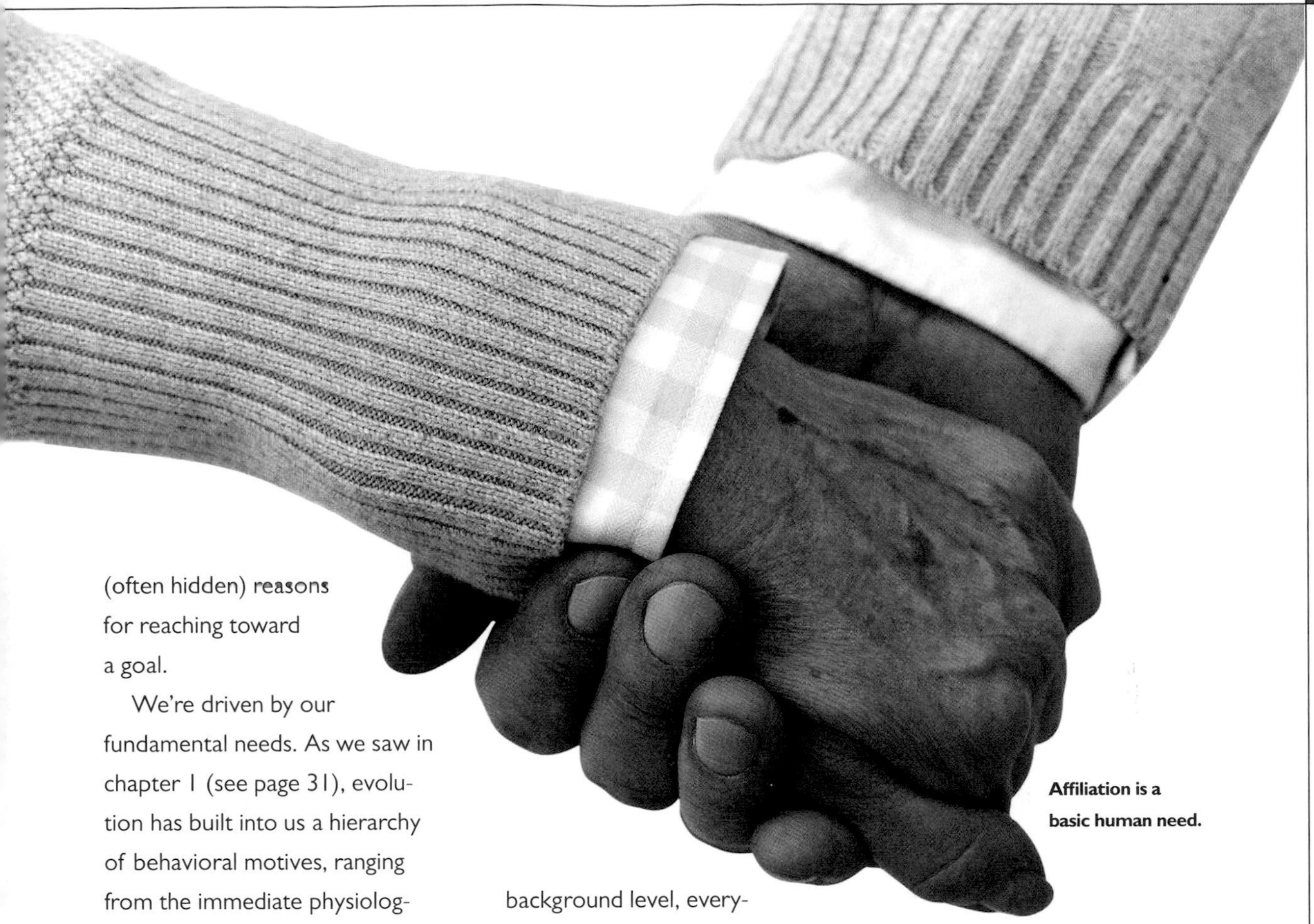

(often hidden) reasons for reaching toward a goal.

We're driven by our fundamental needs. As we saw in chapter 1 (see page 31), evolution has built into us a hierarchy of behavioral motives, ranging from the immediate physiological needs of our body and the need for safety to the desire to acquire a mate and have children. Hunger is a motive that springs from the need for food, and it drives us toward getting that food, or at least toward thinking longingly about eating. Achievement can also be a motive, based on a need to feel successful or competent, and it drives us to take on challenging, but not impossible, activities.

At the most basic, background level, everyone can be said to have two primary motivations. We are prompted either to approach a desired outcome, or to avoid an undesired outcome.

It's not news that we approach pleasure and avoid pain. But the extent to which we pursue the positive or avoid the negative varies from person to person and is correlated with personality traits. More extroverted people are likely to have an approach-oriented temperament; those higher in neuroticism are more likely to have an avoidance-oriented temperament.

Many social choices highlight our approach or avoidance tendency. We know that affiliation is a basic human need and that rejection is literally painful (see chapter 4, page 22). Pursuing affiliation, an approach-oriented person might chat up a stranger at a party. An avoidance-oriented person will veer away from that same stranger, fearful of being rebuffed. Approach or avoidance orientations color

Some people are motivated toward achievement.

Affiliation is a basic human need.

how we interpret social scenes. For instance, participants in one study read an essay about a Saturday night party and then were asked to rewrite the essay from memory. Those with social approach motivations remembered more of the story's positive interactions, while those with an avoidance stance rewrote the story with an accent on the negatives.

Approach and avoidance affect our choices and rewards in close relationships as well. Partners who make sacrifices for a mate because they want to please them or to bolster intimacy report greater satisfaction in their relationships than those who sacrifice to avoid conflict or disappointment.

People motivated by extrinsic goals are less happy than others.

When approach or avoidance comes into play in pursuit of life goals, these motivations are often termed promotion or prevention. Promotion involves moving toward a positive reward; it is a focus on achievement, growth, and accomplishment. A promotion-oriented person might take up running in order to compete in a marathon. Prevention, on the other hand, involves avoiding a negative outcome; the focus here is on safety and security. A prevention-oriented person might begin running in order to avoid getting sick. Promotion-oriented

I got the job!

people approach new relationships by looking for matches; prevention-oriented folks try to avoid mismatches.

Success and failure evoke different emotions in people depending upon their focus. Success makes promotion folks happy and failure makes them sad. Prevention-oriented types feel calm when they're successful and anxious when they fail. Your own reactions to success can tell you which focus you typically have. When you succeed at something, do you feel cheerful, energized, elated? That's characteristic of a promotion focus. If, on the other hand, you feel relieved, relaxed, or less anxious, you have a prevention focus.

Culture affects which outlook you're likely to have. East Asians from more collectivist cultures are more prevention-oriented. They want to avoid failing and disappointing their families and friends. Westerners from more individualistic cultures tend to focus on promotion. They're more likely to pursue success and self-enhancement.

Promotion and prevention approaches can both lead to success or failure. For instance, people with a strong promotion focus are less likely to make errors of omission, while those with a strong prevention focus are less likely to make errors of commission. Think of two people in a classroom: The eager promotion student might raise his hand to answer every question, getting a few answers wrong but many right. The vigilant prevention student raises his hand only when he's confident he has the right answer: He avoids mistakes but doesn't answer as many questions correctly, either. Knowing your own motivations can help you pick problem-solving strategies that will work best for you and keep you more engaged.

Different people are sensitive to different approaches. For instance, in one study students were given a promotion scenario —a student wants to attend an exciting psychology class at 8:30 a.m., so she gets up early—and a prevention scenario—a student wants to take a photography class, so she avoids scheduling a Spanish class at the same time. Later, students with an innate promotion focus remembered the promotion scenario

"The possibilities of human nature have customarily been sold short."

PSYCHOLOGIST ABRAHAM MASLOW

Knowing your own motivations can help you reach your goals.

more clearly, while those with a prevention focus remembered the prevention scenario better.

In general, a promotion focus leads to greater happiness, as well as creativity in problem-solving, even though it can result in greater disappointment at failures. You can consciously activate a promotion focus in yourself or others by setting up positive incentives and cues. The child who gets a trip to the movies as a reward for a good report card, but doesn't lose anything for a poor report card, is being given a promotion focus. She may end up doing better in school than the prevention-motivated girl who receives no reward for good grades but has her phone taken away for a week for poor grades. Aligning a goal with your own values is also more likely to give you a promotion focus. If you want to succeed at a task because it reflects your own core ideals, you'll do better than if you want to succeed out of a sense of obligation.

» Performance and Mastery

Approach and avoidance or promotion and prevention might motivate us toward our goals, but the kinds of goals we seek also shape our identities. In general, we can be said to pursue either performance or mastery goals. You can see this easily in the schoolroom, where one student

“Bore, n.: A person who talks when you wish him to listen.”

SATIRIST AMBROSE BIERCE

might aim to get the top grade on a test, while the student next to him doesn't care about the grade as long as he's confident he mastered the material. Performance goals are about demonstrating competence where others can see it, while mastery goals are about self-satisfaction.

It can be hard to hold on to mastery goals in a world where grades and job titles confer prestige. Nevertheless, mastery goals are more likely to lead to success in the long run. Students with mastery goals (“I want to learn as much as possible from this class”) are more likely to put greater effort into their work, to show more persistence when faced with obstacles, to have more self-control when studying, and to retain information better. They're more likely to ask for help when they need it and also more likely to resist discouragement. For instance, one study asked two groups of participants to answer questions about a psychology text. One group received mastery goals: “You are here to acquire new knowledge that could be useful to you.” The other was primed with performance goals: “You are here to get a good grade on the Multiple Choice Test, to prove your abilities, and to show your competencies.” Then each participant was assigned a remote partner with a scripted response who either agreed or disagreed with him via computer. People who had been given mastery goals were less affected by their partner's discouragement and did better on the task than the performance goal participants, who were thrown off by the negative feedback.

Mastery goals are desirable, but don't fake them if you don't have them. People who endorse mastery goals because that makes other people think better of them perform worse than those who desire mastery for its own sake.

» Multiple Motives

Many students of human behavior have attempted to define

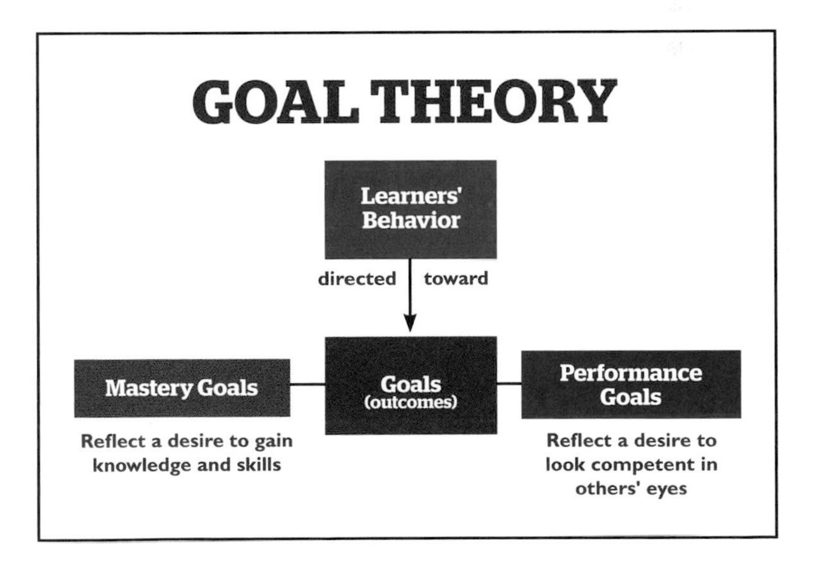

“I can assure you none of us would have expected that degree of similarity.”

PSYCHOLOGIST THOMAS BOUCHARD, DIRECTOR OF THE TWIN STUDY

and name the range of human motives. Most agree that they encompass physical drives as well as emotional ones, and that the strength of these motives varies from person to person. After surveying thousands of people, psychologist Steven Reiss proposed that we all possess 16 basic motivations that organize our lives and that come with intrinsic positive feelings when our needs are met. They are:

- **Power**—the desire to influence. Feeling: efficacy
- **Curiosity**—the desire for knowledge. Feeling: wonder
- **Independence**—the desire to be autonomous. Feeling: freedom
- **Status**—the desire for social standing. Feeling: self-importance
- **Social contact**—the desire for peer companionship. Feeling: fun
- **Vengeance**—the desire to get even. Feeling: vindication
- **Honor**—the desire to obey a traditional moral code. Feeling: loyalty
- **Idealism**—the desire to improve society. Feeling: compassion
- **Physical exercise**—the desire to exercise muscles. Feeling: vitality
- **Romance**—the desire for sex. Feeling: lust
- **Family**—the desire to raise your own children. Feeling: love
- **Order**—the desire to organize. Feeling: stability
- **Eating**—the desire to eat. Feeling: satiation
- **Acceptance**—the desire for approval. Feeling: self-confidence
- **Tranquillity**—the desire to avoid anxiety or fear. Feeling: safety, relaxation
- **Saving**—the desire to collect or be frugal. Feeling: ownership.

These motivations vary in importance from person to person and from one context to another. Each person seeks the balance that suits him or her. Gloria's desire for social contact might be high and not satiated until she has spent hours with her friends. If Daniel's social desires are on the low end, he may be satisfied after about 30 minutes. Too much socialization and he needs time alone to compensate. According to Reiss, these desires organize our lives. We pay attention to those things in our environment that lead to satisfying our desires,

WHO AM I?

"When I'm at home, I think about work. When I'm at work, I think about home. When I'm driving between work and home, I don't know what to think about!"

and tend to ignore those that don't.

» Intrinsic and Extrinsic Motivations

If we're seeking more satisfying lives, we are usually better off if our motives come from inside us (intrinsic motivations) rather than outside us (extrinsic motivations).

The difference between these two motives is highlighted in the common question "Do you live to work, or work to live?" If we work because it brings our lives meaning, our motive is intrinsic. If we work because it brings in money to live, our motive is extrinsic. These labels apply both to the content and the reason for pursuing a goal. A goal's extrinsic content might be money or attractiveness; its extrinsic reason might be to appear successful to your friends. The content of an intrinsic goal might be to have a meaningful life or to forge a close relationship with at least one other person, for the reason that these are your deeply held interests.

Intrinsic goals include self-acceptance, affiliation, community feeling, and physical health. People who pursue intrinsic goals devote more effort to the pursuit and are more successful than those who are motivated by outside factors, such as reward or praise. Intrinsic goals are simply more satisfying. They serve our needs for relatedness, autonomy, and competence. We feel better when we focus on these aspirations.

Extrinsic goals include financial success, image, and popularity.

As rewarding as these seem, people with extrinsic goals are less happy than others. They're less satisfied with their own competence or autonomy and report more competitive and less loving relationships.

However, as most people know, motives are more complicated and situational than that. Sometimes, money does buy happiness, if the money, say, allows us to retire early to volunteer or to paint. Praise and public reward can boost our confidence as we build on our strengths. In any situation, external or internal motivations apply broadly, but our motives are many and are part of a web of desires, emotions, and cultural expectations, all of which shape our behavior.

Self-regulation is one of the most important components of well-being.

THE POWER OF PERSISTENCE

Needs, desires, and motives don't usually line up in a single neat row to carry us to our goals. We all have many things we want or need, some of them conflicting, and not all of them healthy for us. Sometimes achieving a worthy goal means ignoring near-term desires in favor of others that are better for us in the long run.

Consider the marshmallow experiment, originally performed at Stanford University in 1970. This classic study in delayed gratification had some surprising long-term implications. In the original test, experimenters left young children (ages four, five, and six) alone in a room with a marshmallow or other treat. The children were told that they could eat the treat, but if they waited 15 minutes without eating it they could get a second one.

In the face of temptation,

Looking good to others is an extrinsic goal, as are financial success and popularity.

some children ate the marshmallow right away. Most waited at least for a while, while wiggling, covering their eyes, smelling the marshmallow, or tugging on their pigtails as they tried to distract themselves. About one-third lasted long enough to receive the second treat.

Follow-up studies years later showed some fascinating results. Children who were good at delaying gratification in the original tests ended up performing better on the SATs, had better health, and were more popular than others. Brain imaging even showed that those with better self-control had more active prefrontal cortices.

Self-control, also known as self-regulation or simply willpower, is needed when there's a conflict between two impulses—one directed at a momentarily alluring goal, and the other directed at a potentially more valuable, but more distant, target. It is the capacity to override our emotions, thoughts, impulses, or behaviors. Self-regulation is one of the most important components of well-being and is closely linked to the personality trait of conscientiousness. Good self-regulators

THE BIG PICTURE

Should we limit ourselves to traits, needs, and motives when we're studying why people do the things they do? What about basic biology? What about cultural influences?

Researcher Ken Sheldon and colleagues have proposed a big-picture view of human behavior that starts at the atomic level. This multilevel perspective integrates personality and motives into a hierarchy that shows how influences from the chemical to the cultural flow back and forth to affect our actions.

LEVEL OF ANALYSIS (SCIENCE THAT STUDIES IT)

Culture (Sociology, Anthropology)
↓↑
Social Relations (Social Psychology)
↓↑
Personality (Personality/Clinical Psychology)
↓↑
Cognition (Cognitive Psychology)
↓↑
Brain/Nervous System (Neuroscience)
↓↑
Organ Tissues (Medicine, Biology)
↓↑
Cells (Microbiology)
↓↑
Molecules (Chemistry)
↓↑
Atoms (Physics)

Higher levels are supported by lower levels (you can't have cognition without a brain), but lower levels are also affected by higher levels (your cognition changes your brain as well).

Let's say a man decides to get married. At the molecular level, that decision could be based on evolved drives toward mating and reproduction encoded in his genes. Hormones and neurotransmitters at the time of the decision affect the workings of his brain, while cognitive processes weigh the costs and benefits of marriage. His personality and motives will shape his emotional approach to the goal of marriage, as will the mores of his social group and the dictates of his culture at large. The influences flow both ways, with culture, for instance, affecting social rules and motives modifying cognition.

Some levels are more relevant than others. Biology is a primary explanation for our desire to eat dinner, but personality or social relations are more influential when we decide to feed the homeless.

are healthier and do better in school, at work, and in personal relationships. Poor self-regulation is linked to crime and drug and alcohol abuse.

Self-regulation involves some typical processes. You set a standard: a goal or desirable state, such as reaching an ideal weight. You monitor your behavior: Am I sticking to my diet? You deploy your self-regulation, or willpower: This week I'll cut out desserts. And you enlist your motivations: I really want to lose weight so I'll feel better about myself.

» Depletion

You just worked a 12-hour day, focusing on one demanding task after another to meet a tough deadline. When you get home, you just can't face that diet meal. It's time for a pint of Ben & Jerry's on the sofa.

As we all know, self-control can feel like a muscle. After exercise, it's tired and doesn't work well. We feel like we have limited resources when it comes to mental self-regulation, and when they are depleted we wait for them to be replenished. These resources aren't stored in separate stockpiles, either—one

(?) ASK YOURSELF

HOW'S YOUR SELF-CONTROL?

"I can resist everything except temptation," says Oscar Wilde's character Lord Darlington. What about you? How much willpower do you have?

Read the following ten statements and score each one according to the scale beneath it:

1. I have a hard time breaking bad habits.
(1=Very much like me 2=Mostly like me 3=Somewhat like me 4=A little like me 5=Not at all like me)

2. I get distracted easily.
(1=Very much like me 2=Mostly like me 3=Somewhat like me 4=A little like me 5=Not at all like me)

3. I say inappropriate things.
(1=Very much like me 2=Mostly like me 3=Somewhat like me 4=A little like me 5=Not at all like me)

4. I refuse things that are bad for me, even if they are fun.
(1=Not at all like me 2=A little like me 3=Somewhat like me 4=Mostly like me 5=Very much like me)

5. I'm good at resisting temptation.
(1=Not at all like me 2=A little like me 3=Somewhat like me 4=Mostly like me 5=Very much like me)

6. People would say that I have very strong self-discipline.
(1=Not at all like me 2=A little like me 3=Somewhat like me 4=Mostly like me 5=Very much like me)

7. Pleasure and fun sometimes keep me from getting work done.
(1=Very much like me 2=Mostly like me 3=Somewhat like me 4=A little like me 5=Not at all like me)

8. I do things that feel good in the moment but that I regret later on.
(1=Very much like me 2=Mostly like me 3=Somewhat like me 4=A little like me 5=Not at all like me)

9. Sometimes I can't stop myself from doing something, even if I know it is wrong.
(1=Very much like me 2=Mostly like me 3=Somewhat like me 4=A little like me 5=Not at all like me)

10. I often act without thinking through all the alternatives.
(1=Very much like me 2=Mostly like me 3=Somewhat like me 4=A little like me 5=Not at all like me)

Add up your points and divide by 10. The maximum score is 5 (extremely self-controlled) and the minimum is 1 (not at all self-controlled).

“Without distraction, I would have the same thing going round and round in my mind.”

MATHEMATICIAN ANDREW WILES, ON SOLVING A THEOREM

for diet, let's say, and another for relationships. They're all related, a general mental capacity that seems to be exhausted by one kind of challenge and then isn't available for others.

Researchers have tested this in a variety of clever studies. In one, hungry participants were given a bowl of radishes and some chocolates. One group was told it could eat the radishes but not the chocolates (thereby having to self-regulate); the other group could eat either one. Then both groups were given an unsolvable figure-tracing test. The radish-eaters who had to control themselves earlier gave up on the test much sooner than the other group; their ability to discipline themselves was worn out.

In a similar test, subjects who were given two mental tasks—suppressing all thoughts of a white bear (tough!), or solving a simple arithmetic problem (easy!)—were given access to beer afterward. Those who had struggled with the tough mental problem were less able to control themselves and drank more beer.

Depletion of self-control can lead to all sorts of undesirable consequences: sexual infidelity, overindulgence in food and alcohol, and even a tendency to over-share personal information. When their willpower is depleted, people with low self-esteem become even more negative about themselves than usual.

Self-regulation affects more than the self. Healthy relationships and, indeed, social interactions of all sorts depend upon a degree of self-control from the participants. When willpower is depleted, it's hard to resist tempting but unhelpful behavior.

"The Mind Blowing Italian Feast Platter comes with your choice of guilt, regret or defiance."

A study of heterosexual people in committed relationships, for instance, found that when their self-regulation was low they spent more time looking at scantily clad people of the opposite sex in magazine photos. The rule that you shouldn't discuss personal problems when tired seems to hold true, as well. Those with regulation fatigue were more likely to pick arguments, to threaten to leave a relationship, and to hold on to grudges. People who are low in self-control are more likely to cheat and lie, and less likely to say thank you.

Suppressing racial prejudice, in some people, also seems to require a lot of self-control. Researchers compared highly prejudiced white people with those low in prejudice as they interacted with a black participant. Afterward, the more prejudiced subjects did poorly on a test of mental control compared to the less prejudiced folks. Apparently, they were mentally tired after struggling with their own biases.

> **Grit is a key factor in determining success.**

The good news is that this kind of depletion is not inevitable. It has more to do with a shift in motivation than with a basic inability to exert control. Acts of self-regulation are mentally costly, requiring much thought and attention, while their delayed rewards can seem remote and abstract. When we've exerted a lot of willpower already, our motivation shifts away from tiring self-control tasks and toward those that are easier or more immediately rewarding. We can overcome this kind of depletion by bolstering our motivation with the basic belief that we can, in fact, continue to exert control. Studies show that people who are told that willpower is an unlimited resource perform better on demanding tasks than those who are led to think that it can be depleted. In fact, the more that people believe in limited self-control, the poorer their self-regulation is in daily life.

If self-regulation works like a muscle, can we strengthen

it with exercise? Some studies suggest that we can. In fact, physical exercise, because it involves a steady application of willpower, appears to bolster our mental muscles as well. A group of formerly sedentary people who worked out at a gym for two months showed marked improvements on self-regulation tests compared to a control group. Moreover, they also reported that they smoked and drank less, controlled their spending, and even paid more attention to household chores than previously. Another study showed that just two weeks of self-control activities, such as monitoring posture, or keeping track of diet and moods, improved the ability to self-regulate.

Just as it does with physical fatigue, rest can restore depleted resources of self-control. People are more likely to lose self-regulation as the day progresses into evening and to gain it back on Mondays after a weekend of relaxation. But self-control can also be boosted with a shot of positive emotion, such as the enjoyment of a funny movie or a moment of self-affirmation. Strong motivations can counteract willpower fatigue as well. People who are paid for tasks or were told that the results would benefit charity can overcome the effects of fatigue and performed well. Each year, for instance, Penn State students raise millions of dollars for childhood cancer research through a 46-hour no-sitting, no-sleeping dance marathon. Students who typically spend most of their time sitting and watching screens find themselves able to keep moving for two days as they are cheered on by their peers and by recovering children and their parents.

»Grit

When it comes to success in school or at work, perseverance, or grit, is the overlooked partner to intelligence. American psychologist Catherine Cox, who wrote a 1926 dissertation on the mental traits of geniuses, concluded that IQ was only moderately

Self-regulation can be learned and strengthened with practice.

related to success; instead, the qualities of "persistence of motive and effort, confidence in their abilities, and great strength or force of character" were those that predicted lifetime achievement. Charles Darwin had a brilliant insight into the workings of natural selection, but he also spent more than 20 years perfecting the book that made him famous. Thomas Edison tested more than 6,000 different materials for a lightbulb filament before finding the one that worked (supporting his famous quote that genius is "one percent inspiration and ninety-nine percent perspiration").

Grit is a key factor in determining whether a student graduates from high school. A study of 4,813 high school juniors in Chicago public schools found that grit predicted graduation rates, even when controlling for other factors such as perceived motivation, support from parents and teachers, standardized test scores, and demographic variables.

A gritty person, by definition, has high self-control. She is willing to persevere through setbacks to achieve a distant target. However, the gritty individual typically shows a passion for one single overriding goal across a span of years, not necessarily generalized self-control across all spheres in everyday life. Ernest Hemingway, for example, steadily turned out some of literature's most acclaimed novels and short stories while drinking, womanizing, and running with

Exceptional levels of persistence bolstered Diana Nyad as she swam from Cuba to Florida.

the bulls. The gritty person organizes his goals with one single goal at the top of a hierarchy of lesser goals, all in service to the desired end. If an intermediate goal is blocked, the gritty individual finds a way around it.

Grit can overlap with the personality trait of conscientiousness, but it is not the same thing. A conscientious person can be dependable and have excellent self-control without necessarily pursuing a long-term goal for years on end.

» Grit Throughout the Life Span

Grit provides advantages in most arenas. Employers who suffer from high employee turnover rates might want to consider it when hiring, for instance. A study of hundreds of sales representatives in a vacation ownership business found that grittier employees were more likely to remain in their jobs. Grit was a stronger predictor of retention than years in the job.

Grit also predicts success in the military. About half the candidates in the army's grueling Special Operations training session drop out before the end of the 24-day stint. Grittiness is a better predictor of completion than fitness or intelligence. Grit is also a good predictor of which cadets will remain at West Point—better than SAT scores, high school rank, or self-control.

Relationships also benefit from grit, but with an interesting gender twist. In a large study, grit proved to be a better predictor of whether the subject was married and would stay married than standard personality traits. However, this was true only for men. In women, there was no correlation between grit and marital status.

Can we build grittiness in ourselves or our children? The jury is still out on this one, but some researchers believe that cultivating a growth-oriented mind-set is one key to persistence. People with a fixed mind-set believe that their abilities and outcomes are

PRACTICE MAKES PERFECT

It's an old joke with a valid message:

"How do you get to Carnegie Hall?"

"Practice, practice, practice."

Nowhere is this lesson more in evidence than in the cutthroat competition known as the Scripps National Spelling Bee. During the 2006 contest, psychologists tracked 190 of the finalists to see which of three kinds of training regimens was most successful: leisure-time activities such as reading or games; being quizzed by parents or computer; or deliberate solo practice of word spellings and origins.

The winner—deliberate solo practice. Although solitary drilling was the least enjoyable of the three methods, children who used it the most performed better at the spelling bee. In the three years before the contest, each one spent hundreds of hours researching words.

These contestants, the researchers found, were the most naturally persistent. Asked on a survey to agree or disagree with statements such as "Setbacks don't discourage me," the students who agreed most strongly were the most likely to plug away at practice.

The final word in the 2006 bee? "Ursprache" (a reconstructed protolanguage).

innate and unchangeable. They might respond to setbacks by thinking, "I'm a failure," or "He'll never let me win." Those with a growth mind-set believe that they can learn from their mistakes and do better next time. For this reason, many psychologists recommend praising children not for their innate intelligence ("Oh, you're so smart") but for their persistence ("I'm proud of you for working so hard").

Finally, researcher Angela Duckworth points out that the typical educational system, which

Conscientious people study for tests, make their beds, and plan ahead.

insists that students pursue a broad range of courses and activities, may not nurture the true achievers among them. "To paraphrase Benjamin Franklin," she writes, "the goal of an education is not just to learn a little about a lot but also a lot about a little."

Much of who we are arises from our biological history. Our personalities, our drives, our approaches to problems and goals, are all to some extent inborn. This doesn't mean that they are unchangeable. Once we recognize our own characteristic traits, our typical motives, our ability to control our behavior, we can strengthen the positive and dampen the negative. We won't all climb El Capitan, or want to, but we can all harness our drives and persistence to reach the peak of our own particular mountain.

Parents can foster grit in their children by rewarding persistence.

The MIND &

» PART THREE

HEALTH

The complex mental processes that keep us safe and help us connect to others can also go astray. Few people are free from anxiety, emotional turmoil, or self-doubt. Sometimes, these fears can derail a productive life. In these final chapters, we'll examine the ways in which simple and practical changes can put us back on the path to health. Psychologists are recognizing that natural sources of strength, such as spirituality, as well as simple interventions that refocus our minds on gratitude and acceptance, can lead us to a state of positive, flourishing well-being.

» CHAPTER SEVEN «

HEALTH, ILLNESS & THE STATES IN BETWEEN

Football star Ricky Williams should have had every reason to be confident. A Heisman Trophy winner at the University of Texas, he went on to play professional football with the New Orleans Saints and then the Miami Dolphins. Yet fans and the press began to deride him as an oddball.

He avoided other people: teammates, onlookers, and reporters. He conducted interviews with his helmet on and his tinted visor down. He repeatedly failed drug tests.

"When I was [drafted to] New Orleans, it got to the point where I didn't want to leave my house. I didn't want to go anywhere. I didn't want to go to the grocery store. I didn't go out on dates," the football star said. Until he went to therapy, Williams himself did not know why he behaved the way he did. Once in treatment, he learned that he had social anxiety disorder, a condition that left him intensely afraid of being scrutinized by others. It was welcome news. "After I was diagnosed with social anxiety disorder, I felt immense relief because it meant that there was a name for my suffering. I wasn't crazy or weird, like I thought for so many years," said Williams.

Williams wasn't crazy or weird. He wasn't completely healthy, either. He was like many people—struggling with a disorder that hampered but didn't destroy his life. And like many people, he found a diagnosis, treatment, and a path back to well-being. In the 21st century, psychologists have built upon a foundation of biological and social research to develop practical approaches to improving mental health. They range from cognitive therapies to simple, at-home exercises that anyone can do to achieve a happier, more fulfilling life.

LANGUISHING OR FLOURISHING?

If mental health is defined as the absence of serious mental illness, most of us are

ASK YOURSELF

ARE YOU LANGUISHING OR FLOURISHING?

Mental health means more than avoiding serious disorders. It also means embracing positive, enriching beliefs, habits, and relationships—flourishing rather than languishing. From this perspective, are you in the peak of mental health? See where you stand by answering yes or no to the questions below (adapted from a survey by Emory University psychologist Corey Keyes).

» ARE YOU:

1. **Regularly cheerful, happy, calm, satisfied, and full of life?**
2. **Feeling happy or satisfied with life overall?**
3. **Holding positive attitudes toward yourself and your past life and accepting varied aspects of yourself?**
4. **Feeling positive toward others, while accepting their differences and complexity?**
5. **Showing insight into your own potential and feeling open to new and challenging experiences?**
6. **Believing that people, social groups, and society can evolve and grow positively?**
7. **Holding goals and beliefs that affirm a sense of direction in life and feeling that life has a purpose and meaning?**
8. **Feeling that your life is useful to society and valued by others?**
9. **Showing the ability to manage a complex environment and mold your environment to your needs?**
10. **Interested in society and social life and feeling that society and culture are predictable and meaningful?**
11. **Able to be guided by your own socially accepted internal standards and able to avoid unsavory social pressures?**
12. **Experiencing warm, satisfying personal relationships and feeling capable of empathy and intimacy?**
13. **Feeling as though you belong to a supportive community?**

The higher the number of positive responses, the higher you place in the range from languishing to flourishing: More than six "yes" answers indicates a positive state of health.

just fine. Three out of four American adults make their way through any given year without experiencing any mental disorders. Each year, sixteen out of seventeen live without serious illnesses such as depression or schizophrenia.

For a long time, mental health was defined in just this way—the absence of illness—and then ignored. Understandably, psychologists concentrated on finding treatments for the serious disorders that disrupt so many lives. But now we recognize that mental health is more than just the absence of disease. It is the presence of something healthy: a productive state of emotional and social well-being. The 2000 Surgeon General's report defines mental health as "a state of successful performance of mental function, resulting in productive activities, fulfilling relationships with people, and the ability to adapt to change and to cope with adversity."

Just as illness can fall on a spectrum from mild to severe, mental health can span a continuum ranging from languishing to moderately healthy to flourishing. And just as mental illness is defined by its symptoms, so too is mental health. Symptoms such as a positive attitude toward yourself or a thriving social life are indications of well-being (see sidebar, opposite). The more positive emotions, beliefs, and behaviors you have in your life, the more your mental health is approaching optimal level. Well-being is an attainable goal for most of us, as long as we know ourselves and the resources we can draw on.

How many of us languish, flourish, or coast along in between? National surveys indicate that roughly 17 percent of Americans are languishing, feeling relatively unhappy and struggling to connect with life. A majority is in the middle. About 18 percent can be called flourishing: happy, positive, productive folks. These lucky people are not only happier, they are physically healthier than others. They have fewer chronic diseases, fewer doctor visits, and fewer missed workdays than anyone else on the spectrum. They feel less helpless and more purposeful than others and report greater closeness to family and friends.

These conditions of mental health exist independently from, but are correlated to, mental illness. In other words, you can have symptoms of mental health and symptoms of mental illness at the same time. In general, though, the more you are languishing, the higher the probability that you also have some form of mental illness. Almost a third of mentally unhealthy, languishing people have had a major depressive episode, compared to only 5 percent of those who flourish. About 16 percent of languishers have a panic disorder, compared to less than one percent of the flourishers.

If you have a mental illness, but also strong symptoms of mental health, you can fare relatively well. People with a mental illness who ranked in the middle or high end of the separate positive mental health scale reported better physical health and fewer workdays missed than those at

People can have symptoms of mental illness and mental health at the same time.

the languishing end of the mental health scale. People hit with the double whammy of a mental illness plus a low score in mental health, not surprisingly, fared worse than any other group.

SHADOW SYNDROMES

Kate is down in the dumps. She's sad, listless, and has trouble taking pleasure in many things. On the other hand, she still gets her work done, eats well, and shows up for her exercise class. Is she clinically depressed?

Daniel is a worrier. He frets about his health, his work, and his family. Sometimes his concerns keep him up at night, unable to escape his thoughts. Even so, he enjoys his job, his friends, and seeing new places. Does he have an anxiety disorder?

It's possible that neither of these individuals would meet the diagnostic criteria for a mental illness. Nevertheless, they have real and serious problems that might benefit from some kind of mental health support. Many people are like them, struggling with some form of everyday mental distress, but not ill enough to meet the standards of a defined disorder. These shadow syndromes can encompass a range of problematic thinking and behaviors that leave a person vulnerable to more serious mental illnesses.

» Avoidance

We know that we can view our thoughts and behaviors in many ways, one of them being that of approach or avoidance motivation. People with an approach orientation focus on gains, growth, and achieving positive outcomes. Those with an avoidance outlook concentrate on losses, safety, and avoiding failure. One orientation is not necessarily better than the other, but the avoidance perspective can get out of hand.

Avoiding the things you fear is natural, but it can also do you harm. The advice "face your

FOCUS

DIAGNOSIS & DEBATE

THE ANSWERS

In print since 1952, the *Diagnostic and Statistical Manual of Mental Disorders* is now in its fifth edition (DSM-V). It is the official guide, almost 1,000 pages long, to diagnosing mental illnesses ranging from agoraphobia to voyeuristic disorder. For each disorder, the DSM lists the symptoms that should be present (and those that should not be), describes the illness, and classifies it with a code that practitioners use for billing, among other things.

The DSM-V has come under fire from many directions for a variety of perceived faults. Common criticisms:

- The criteria are too vague. For example, there are 636,120 ways to have post-traumatic stress disorder (PTSD).
- The criteria are arbitrary and unscientific.
- The criteria are too inclusive, turning routine problems into disorders. For instance, the current edition includes disruptive mood dysregulation disorder (DMDD): recurrent temper tantrums in children.
- The disorders may reflect the contributors' pet causes, possibly stemming from their own areas of research or those of their financial backers.

Psychologists and others have long debated just how to define mental disorders, as changeable, complex, and interconnected as they are. The current manual seems unlikely to end that debate.

fears" is sound. The more you avoid the thing you fear, the more fearful you become. The more fearful you become, the more you avoid the frightening thing, and on and on. For example, if you're afraid of driving over bridges, at first you might avoid crossing just the biggest ones. But the very act of avoidance keeps the fear in the forefront of your mind, allowing you time to imagine what might happen if the bridge collapsed or if you passed out while driving over it. Not only do you become more likely to avoid big bridges, but also the fear starts to generalize to smaller bridges (what if *this* bridge collapses?). You don't notice signs of safety because you're too busy attending to possible threats. Your mind becomes trapped in worst-case scenarios and you learn that driving over bridges is in fact a terrifying experience—for you.

You can defuse your fears by facing them head-on, rather than avoiding them.

When you avoid not just places or events, but also your own unpleasant thoughts, feelings, and internal sensations, it's known as experiential avoidance. None of us seeks out disagreeable feelings, but when avoidance becomes a series of regular, deliberate attempts to control or escape natural thoughts or sensations, then it becomes counterproductive. Your thinking can become a disordered process in which you devote enormous time, effort, and energy to controlling or struggling with unwanted internal events. The struggle gets in the way of moving you toward your goals and distances you from the pleasures of daily life. Eventually, the act of avoidance is more psychologically damaging than simply experiencing the unpleasant thoughts.

Let's say you're looking for a new job. It's useful to suppress some feelings of anxiety during a job interview. It's counterproductive, though, when you start to avoid interviews because they're too stressful, or when you begin to label yourself as too shy or too fearful to ever hold a responsible job.

Trying to block unwanted thoughts, feelings, or desires can backfire. Take a relatively benign example: contemplating chocolate. In one study, researchers asked participants to record their thoughts. One-third were

TRY IT

DON'T EVEN THINK ABOUT IT

Dr. Daniel Wegner conducted a classic experiment about thought avoidance in the 1980s. Try out one version yourself:

Read the following paragraph. While you read, consciously avoid thinking about a white bear.

"Like so many other rituals, the origins of the handshake are obscure. In ancient Babylonia, a ruler would take the hand of a sacred statue to symbolize the transfer of divine power to human hands. However, the modern purpose of the handshake is quite clear: It's meant to convey open, safe greetings between two people. Actually, make that 'between two men.' We do know that handshaking was used several hundred years ago in England to demonstrate that no arms were being carried. Women rarely carried (or were allowed to carry) weapons, and that's probably why the handshake was not a common greeting for women until recently."

Were you able to block any thoughts of a white bear from your mind while reading? If not, how often do you estimate you thought about it? In similar experiments, the great majority of readers were aware of the bear padding around in their brains the whole time. Consciously not thinking about the bear is, in effect, the same thing as thinking about the bear.

asked to think about chocolate; one-third to suppress thoughts of chocolate; and one-third simply to record any kinds of thoughts. Later, all of them were asked to rate some chocolates according to taste. The ratings were not the point, however: The researchers found that the people who tried to suppress thoughts about chocolate not only thought more about chocolate, but ate more of it than the other two groups. Attempting to avoid the sweet temptation simply made them want it more.

»Harmful Perspectives

The ability to perceive an approaching threat, a thoughtful approach to problems, and a dedication to high standards are

Resistance is futile.

Useful traits such as perfectionism can become maladaptive if overdone.

all useful adaptations to life, in moderation. Sometimes, these thought patterns become maladaptive. They can invade our lives and increase our vulnerability to psychological disorders.

One such maladaptive mental trap goes by the curious name of cognitive looming. This is the tendency to see any threat or worry as rapidly approaching, or looming toward you. Cognitive looming often plays a part in phobias. Imagine, for instance, a big black spider sitting right in front of you. You can see every detail: its long skinny legs, its eight beady eyes. Now imagine it scuttling toward you. It's getting bigger. You can hear its skittering legs and see that it's about to jump onto your hand. Which image is scarier: the motionless arachnid, or the approaching one? People with phobias tend to see the objects of their fear as constantly moving toward them. Anxious people may feel this way all the time. They see a host of everyday worries as continuously advancing on them, gaining speed, increasing their fear and distress.

Cognitive fusion is another pitfall. The "fusion" here is the fusion of your thoughts and reality. When you don't recognize passing thoughts and feelings as just that, and believe that whatever you're feeling is the objective truth, you're trapped in cognitive fusion. "I'm never going to be happy"; "People hate me"; or "I'm terrible at socializing" are all examples of cognitive fusion. This problematic thinking pattern has real behavioral consequences, as the person withdraws from others

"Cognitive looming" means the things you fear seem to be always approaching.

or stops trying to make positive changes. You can also fuse with a perceived role in life. Women who have devoted years to raising young children may come to see themselves as mothers only, forgetting other identities they once may have had, such as professionals or free spirits who enjoyed adventure. Past identities can become fused into present perceptions. If you were the black sheep of the family, or the loud partier in college, that role may be a bad fit for your current life, but one that's hard to discard.

Fortunately, there are therapies that address these disabling thoughts. They may start simply by asking you to distance yourself from the thought by identifying it as such: "I'm having the thought that I'm never going to be happy." Labeling your belief as the passing thought that it is gives you distance and perspective, opening up room to make changes. Therapists may also take their clients through verbal exercises that help them understand that thoughts and words are not the same as the things they define. They might, for instance, describe a chair down to the last detail and then ask the client if he could sit in it. The point is that a mental experience is not the same as a real-life experience, but sometimes we need to explicitly work through that thought to take it in. "Defusing" therapies such as these (see chapter 9, page 266) have been shown to ease workplace stress and a wide variety of disorders from panic disorder to social anxiety.

Avoidance keeps a fear at the forefront of your mind.

Rumination, another thought pattern, can be productive or destructive. Imagine this scenario: Before leaving for the day to join friends for dinner, you open one last email. Unexpectedly, there's a note from your boss. She tells you that you've made a serious error at work. No explanation needed, but you need to turn things around in the coming weeks. What do you do?

1) You close your laptop, head out to dinner, and enjoy yourself.

You'll figure out your strategy tomorrow at work.

Or

2) You go to dinner, but spend the evening obsessing over the email, your boss, and how hard it will be to fix things.

Ruminating over troubles is natural, but when you go over and over the same thoughts, focusing on your distress, it becomes a destructive habit. Brooding about a stressful experience keeps the stress active and your mind focused on unproductive thoughts. Studies show that rumination leads to a decrease in performance, as the brooding person's attention is diverted from other tasks at hand. It's also linked to a greater vulnerability to depression and anxiety disorders, post-traumatic stress, binge drinking, and eating disorders.

It is possible to ruminate productively. People who ruminate about how to improve their performance, but don't mull over their troubles, are more likely to solve their problems and move on. But for most people rumination is a time-waster that needs to be corralled. Psychologists recommend some commonsense tactics. Take a clear look at your fear and identify it. What's the worst-case scenario? Can you deal with it? (Usually, it's not so bad, and yes, you can.) Identify what you can change and let go of what you can't control. Take a break for exercise, mindfulness exercises, and other stress-reducing activities.

Perfectionism is another attitude that can be a plus or a minus. Across cultures, it's associated with greater hope, higher standards, and better organization. However, perfectionists who berate themselves for failing to meet their own high standards can bury themselves in negativity. Self-critical perfectionism is associated with depression, feelings of guilt and shame, procrastination, and substance abuse.

Destructive perfectionism has three varieties:

- **Self-oriented perfectionism:** setting unrealistically high standards for your performance and then judging the results harshly.
- **Other-oriented perfectionism:** setting unrealistic standards for others. This can lead to blame, hostility, and a lack of trust toward other people.
- **Socially prescribed perfectionism:** the belief that other people are holding you to high standards, leading to fears of their criticism. Folks with this kind of fear expend a lot of energy trying to avoid disapproval.

If you temper perfectionism with flexibility, it can be a positive trait. People who strive for greatness, while realizing that they'll make mistakes along the way, will often flourish.

»Emotional Control

A tired child is whining to his father, begging to be picked up. The exasperated parent snaps and yells at the child. The child wails. The father feels guilty.

Friends gather at a restaurant to celebrate an engagement: One of them runs from the table in tears, desperately jealous.

These unfortunate scenes spring from a failure in emotion regulation, the process by which we tailor our emotions, and our expression of them, to suit the situation. Our emotions guide us at each step of the way as we notice, assess, and respond to situations for better or worse. The stressed-out parent can control his reactions at various steps in the process: when the child begins to whine, when the father thinks about what to do, when he responds to the child—and when he either yells at the child,

When ordinary situations send us out of control, we're failing at emotional regulation.

TRY IT

ACHIEVING "GOOD ENOUGH"

If you're the safety chief at a nuclear plant, then perfectionism may be necessary to your job. For most of us, however, it's fine to simply do our best. Here are some ways to cope with damaging perfectionism.

Make sure you have realistic goals. Are they specific, measurable, and attainable?

Look at the big picture. Maybe you didn't run ten miles this week—but overall you've been running steadily and well.

Listen to yourself. Be careful of judgmental terms such as "should," "shouldn't," or "must."

Recognize specific behaviors and beliefs. Do you check your work over and over? Do you avoid new challenges because you're sure others will judge you for your failures?

Confront those behaviors. Set yourself specific tasks: Tonight I'll turn off the computer and forget about work until tomorrow.

Check the results. Did disaster result from your new behavior? Did you feel better or worse?

"SHOULD" "SHOULDN'T" "MUST"?

62
МЕГАФОН
СОГАЗ
37
MX3 SUPREME
ERIEL
ПРОМБАНК
52
BAUER

"Worry is interest paid on **trouble before it falls due.**"

WRITER W. R. LANGE

changing the situation for the worse, or decides to control his anger and soothe his son, creating a new and calmer situation.

We all do this. We modify our feelings to smooth social situations, ease stress, and reduce emotional vulnerability. Most of us regulate not only how we feel, but also how we act. We curb the impulse to snap at a cranky toddler. We put on a happy face for a fortunate friend. And it's not just negative emotions that need to come under control. People who were asked about times in the last two weeks when they regulated their emotions most commonly listed anger, sadness, and anxiety as the targeted emotions—but they also mentioned happiness and romantic attraction as feelings that needed to be reined in.

Effective, appropriate emotion regulation is a healthy habit. Poor regulation is a defining feature of many disorders, including major depressive disorder, generalized anxiety disorder, bipolar disorder, social anxiety disorders, and substance abuse. On the other hand, emotional control can be overdone if it results in avoiding or attempting to suppress your own emotions entirely. More helpful are strategies such as reappraisal, in which you distance yourself from the situation and put it into context. That whining child is just tired, and so are you. You'll both fare better after some rest.

Emotional control is not a priority In the sport of hockey.

In stressful moments, take a time-out and reappraise the situation.

The Spectrum of Anger

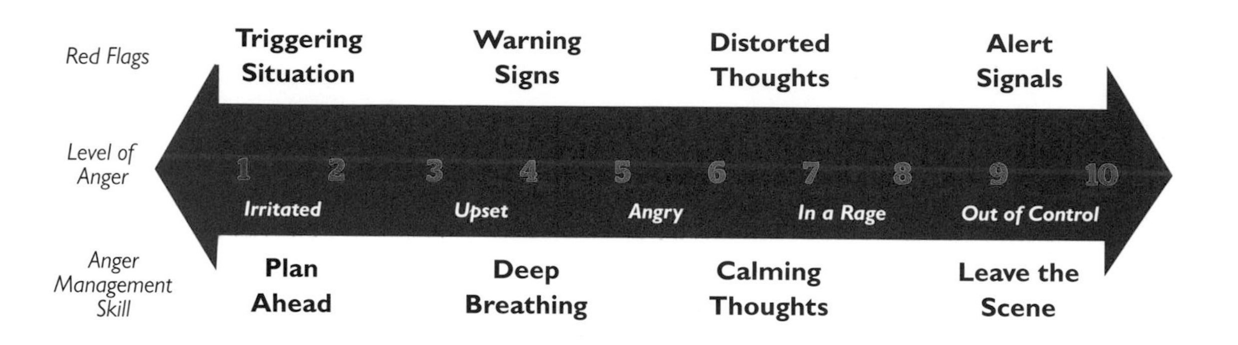

Reappraisal also lets you reframe a situation and give it a different emotional meaning. If you fail a test, for instance, your first reaction might be to see yourself as a failure, incapable of passing the course. However, you can change your emotional trajectory by reappraising the situation. Now you know what you need to study, and you'll rise to the challenge and learn new things. With this new attitude, you can go back and actually change your situation for the better. People who are good at reappraising have been found to be more optimistic and to have better social relations than others.

If the ability to control your emotions is generally a good thing, so too is the ability to tolerate distress. People who often find themselves saying they "can't bear" or "can't handle" guilt, anger, anxiety, or other unpleasant emotions may have a low threshold for emotional pain. As a group, they are more likely to suffer from depression, anxiety, and risky, addictive behaviors. Once addicted, they have a harder time quitting and are more likely to relapse when they do. (How do psychologists measure distress tolerance without torturing their subjects? They time how long they can hold their breath, or how long they can hold their hands in cold water.)

Self-awareness is one key to raising your threshold for emotional pain. Avoiding distress allows it to build up; noticing it helps you respond appropriately and think of solutions. Distress can also be a valuable signal. The marathon runner who races through increasing ankle pain may be setting himself up for permanent injury. The goal is to recognize the distress, understand what it's telling you, and decide whether it's healthier to persist or to quit.

Psychological rigidity is another dysfunctional approach to daily life. Let's say you've learned to get your way by deferring to others, being polite, and backing away from angry feelings. This may succeed much of the time, but when it comes to dealing with a troubled relationship, say, or getting care from an indifferent doctor, you may be better off enlisting a modicum of anger and confronting the other person directly.

Flexibility, the ability to draw on different approaches and emotions as you pursue your goals, is a healthy skill. So is the capacity for balancing life's domains. The conscientious employee who can't switch off on the weekend, or the worrier who is too focused on daily troubles to think about long-term goals, is caught in a limited life. If you can switch your focus when needed from long-term goals to the here-and-now, from work to play, from the big picture to details, and back again, your life will be richer and more meaningful (see chapter 9, page 266). Having a rigid attitude, on the other hand, can lead to trouble; it's associated with higher stress levels and greater difficulty in recovering from disorders such as social anxiety or borderline personality disorder.

CROSSING THE THRESHOLD

Many people struggle with worries, poorly controlled emotions, inflexible attitudes, and the like, but few cross the line into full-blown psychological disorders. What pushes a person over the edge? Why can four soldiers suffer through a wartime explosion, but only one develop post-traumatic stress disorder? Reduced to a simple

Being able to balance the demands of daily life is a hallmark of psychological flexibility.

formula, the answer is stress + vulnerability = disorder.

» Stress

The first part of the equation, stress, refers to the bad stuff that life throws at each one of us. The death of a loved one, the end of a romance, abuse from a parent, witnessing a terrible accident—all these and more are serious stressors.

You can also be your own source of stress. Having a negative, glass-half-empty response to a stressful event can add weight to that event. Sometimes we precipitate painful events ourselves. A woman whose mother has just died might withdraw into depression and pull away from her spouse. This emotional withdrawal could lead to separation or divorce, another source of stress.

» Vulnerability

Vulnerability is the second part of the equation. It is what you bring to the table, the set of factors that makes you susceptible to developing a disorder. These factors can be biological, such as a genetic susceptibility to a disorder like anxiety or depression. They also include our biases and expectations, our existing dark emotions, or patterns of dysfunctional behavior in our lives.

Typically the two factors, stress and vulnerability, work hand in hand, but one may play a much larger role than the other. In some cases, a single factor is all it takes. For example, schizophrenia is strongly linked to genetics. A person can develop the disease without any major stressors simply because he was dealt a bad genetic card. Similarly, a serious trauma such as a rape or a car accident can on its own lead to an illness such as post-traumatic stress disorder, even if the sufferer has little vulnerability to the malady.

ANXIETY

Mental illnesses take many disparate forms, but the most common are the anxiety disorders. In any given year, about 18 percent of Americans will suffer from serious anxiety. Anxiety disorders have a multitude of symptoms, ranging from a fear of strangers to compulsive hand washing, but they have some key traits in common. In general, anxiety is a negative mood state accompanied by physical symptoms such as a pounding heart, tense muscles, a sense of unease, and worry about the future. Seriously anxious people feel helpless against threats, and yet are hypervigilant,

If you're vulnerable to disorders, a little stress can push you over the edge.

always on the lookout for the thing they fear.

Like some other disorders, anxiety represents an evolutionary adaptation that has spun out of control. Our early ancestors were more likely to survive if they were apprehensive about predators, strangers, storms, or venomous animals. Today, anxiety is useful if it prompts us to study for that test or check to make sure the door is locked before going out. It becomes a problem only when it becomes overwhelming and disabling.

People with anxiety disorders can become paralyzed with fear; they will avoid anxious situations to the point of not leaving the house or refusing to meet new people.

FOCUS

KAYAK ANGST

In 1963, psychiatrist Zachary Gussow noted a strange disorder among Inuit seal hunters in West Greenland. Out alone in his kayak, paddling through glassy waters, the hunter suddenly becomes dizzy and disoriented. He holds as still as he can, sweating and trembling, trying to control himself even as he becomes convinced that the kayak is filling with water and sinking. Paralyzed with terror, the hunter might believe he is being pursued from beneath or behind. If he can master himself enough to start paddling, the symptoms abate when he reaches shore. For many of these hunters, the attacks return repeatedly until they no longer venture onto the water.

Dubbed "kayak angst," this kind of episode is now recognized as a form of panic attack, albeit one tailored to a unique environment. Its symptoms—a sudden onset, dizziness, racing heart, sweating, intense fear, and an avoidance of the place where a previous attack happened—are typical of panic attacks the world around. These kinds of attacks, which last only a few minutes but feel as though they go on much longer, affect about 6 million Americans. Some lucky individuals experience only one in a lifetime. Others endure them repeatedly, to the point where the attacks restrict their jobs and activities.

Like some other anxiety-related disorders, panic attacks may involve the brain's limbic system, a region connected to emotional stimulus and fear. Panic attacks are frightening but quite treatable, usually with a combination of therapy and medication.

»A State of Dread

Everyone has spells of anxiety: Will I meet this deadline? Can I pay this month's bills? What will I do about my aging parents? Usually these worries ebb and flow, fading away when the problem is solved or a cheerful occasion intervenes. But for some people, the anxiety never ends. They live in a continual state of dread—tense, shaky, restless, yet tired. "What if" scenarios haunt them: What if I have cancer? What if my car slides off the road in the rain? They have trouble sleeping at night and concentrating during the day. Worst of all, they can't point to any specific cause that they can cure or avoid.

People who live like this for six months or more might end up diagnosed with Generalized Anxiety Disorder (GAD). Much worse than routine anxiety, GAD can be disabling, affecting work, relationships, and social life. It's one of the more common anxiety disorders, affecting almost 6 percent

"What if you could talk to the boy who scooped out your insides right now. What would you say?"

of the population at some point in a lifetime. Many of its sufferers have depression as well.

People who develop GAD are highly intolerant of uncertainty. Any ambiguous situation sets off a round of worrying, which they hope will either help them deal with a fearful result when it happens or maybe prevent that fearful result from happening at all. Problems look like threats, but the anxious folks aren't confident that they can deal with the threat.

This constant worry about minor situations can be a way to distract oneself from more serious issues. People with GAD are more likely than others to endorse the statement, "Worrying about most things I worry about is a way to distract myself from worrying about even more emotional things." If you're busy fretting about your project at work, it might be your way of avoiding thinking about a troubled relationship at home. It's a coping strategy, useful in the short term, but harmful in the long run because it ends up restricting the sufferer's behavior and emotions.

About 6 million Americans experience panic attacks.

"I am crap at parties. I'll sit in the corner and find one person to talk to."

ACTRESS KEIRA KNIGHTLEY

Like other anxiety illnesses, generalized anxiety disorder can be self-reinforcing. A mother fretting that her daughter will be kidnapped during a sleepover might call her over and over. In the morning the girl is safe and sound, but her mother is worse. Her irrational behavior has seemingly been rewarded with a positive outcome, which makes her more likely to do it again.

»Don't Look at Me

If the situations you dread are social ones—when you go far out of your way to avoid attention—you might be experiencing social anxiety, the disorder that afflicted football player Ricky Williams. Like other anxiety disorders, it's not so much the quality as the quantity of symptoms that turn social anxiety into an illness. Most of us are at least a little apprehensive when we go on a first date or have to give a speech. But those with the disorder take it further: They're afraid of multiple social situations—such as talking to a stranger or eating in front of other people—for at least six months. All of these situations boil down to one core anxiety: the fear that other people will see your inner flaws and that they will judge and reject you.

Despite its name, social anxiety is not about other people: It's about yourself. Socially anxious people are highly self-critical. They want to make a good impression, but they don't believe they will because they're sure that they are too boring, unintelligent, or awkward. They're convinced that they have more flaws than the ordinary person. Coupled with that are unreasonably high standards for their own behavior: "I have to sound really smart" or "be completely charming" or "always know the answers." They follow up these beliefs with the assumption that any awkwardness will lead others to judge them harshly: "If I blush, he'll think I'm a complete fool."

People with social anxiety strive to avoid attention.

FOCUS

THE SCARY SOCIAL SCENE

Are you knock-kneed when giving a speech? Tensed up at exam time? You're not alone. People with and without social anxiety disorder (SAD) share the same set of common social fears—but they are far more prevalent among those with SAD. For example, the vast majority of people with SAD fear public speaking, meeting new people, and talking to those in authority, compared to a fifth or less of the overall population. The chart on page 224 shows the most common social fears and their prevalence over a lifetime.

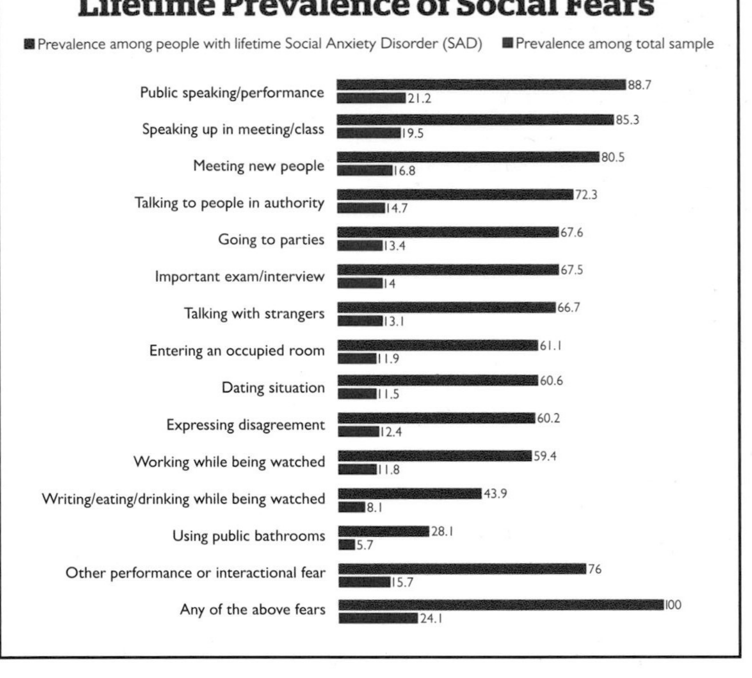

This kind of anxiety seems to have a numbing emotional affect. One study that asked participants to report on their feelings at random times during the day (through portable devices) found that socially anxious people reported less intense positive feelings than others, even during sexual encounters.

People with social anxiety avoid even positive attention. They discount praise when they receive it and squirm away from being commended in front of a group. Why? One theory: People are social animals descended from social animals. Our ancestors may have evolved behavior that wouldn't threaten the status quo. Challenging a higher-status member of your group through your strength or achievements could have led to a world of hurt. It may have been safer to lie low, not make waves, and have lots of babies.

»Post-Traumatic Stress Disorder

Englishwoman Lisa French boarded a bus in London's Tavistock Square on July 7, 2005, and woke up, eardrums blown out, to find the bus destroyed around her by a terrorist bomb. It's hardly surprising that she developed an intense fear of any public transportation and couldn't bring herself even to get on a train. Her post-traumatic stress disorder (PTSD), which later yielded to therapy, was in its way a logical response to a dangerous experience.

Evolution has shaped our brains to remember and avoid dangerous situations. In PTSD, however, people don't just

"There cannot be a crisis next week. My schedule is already full."

FORMER SECRETARY OF STATE HENRY KISSINGER

remember the danger, they relive it, experiencing the fear that went with it as if it were happening for the first time. Memories of the trauma don't fade, but become disabling.

Symptoms of PTSD include:

- **A traumatic event in your past.** You experienced in some way a frightening event that threatened death or injury. Accidents, abuse, earthquakes or storms, or the death of a loved one are common instigators of PTSD. You don't have to have been the person involved—you can have heard about it or witnessed it.
- **Reexperiencing the event.** Flashbacks, nightmares, and unwanted memories come back over and over. Fear and the symptoms of fear—a racing heart, sweaty palms, or hyperventilation—accompany the memories.
- **Avoidance.** You try to avoid anything that reminds you of the trauma or the feelings associated with it.
- **Unhelpful changes in thought and mood.** These include feelings of alienation, negative attitudes toward yourself and the world, and decreased interest in once-pleasurable activities.
- **Agitation.** You might be easily startled or irritated, find

Eleven percent of New Yorkers reported PTSD in the months after 9/11.

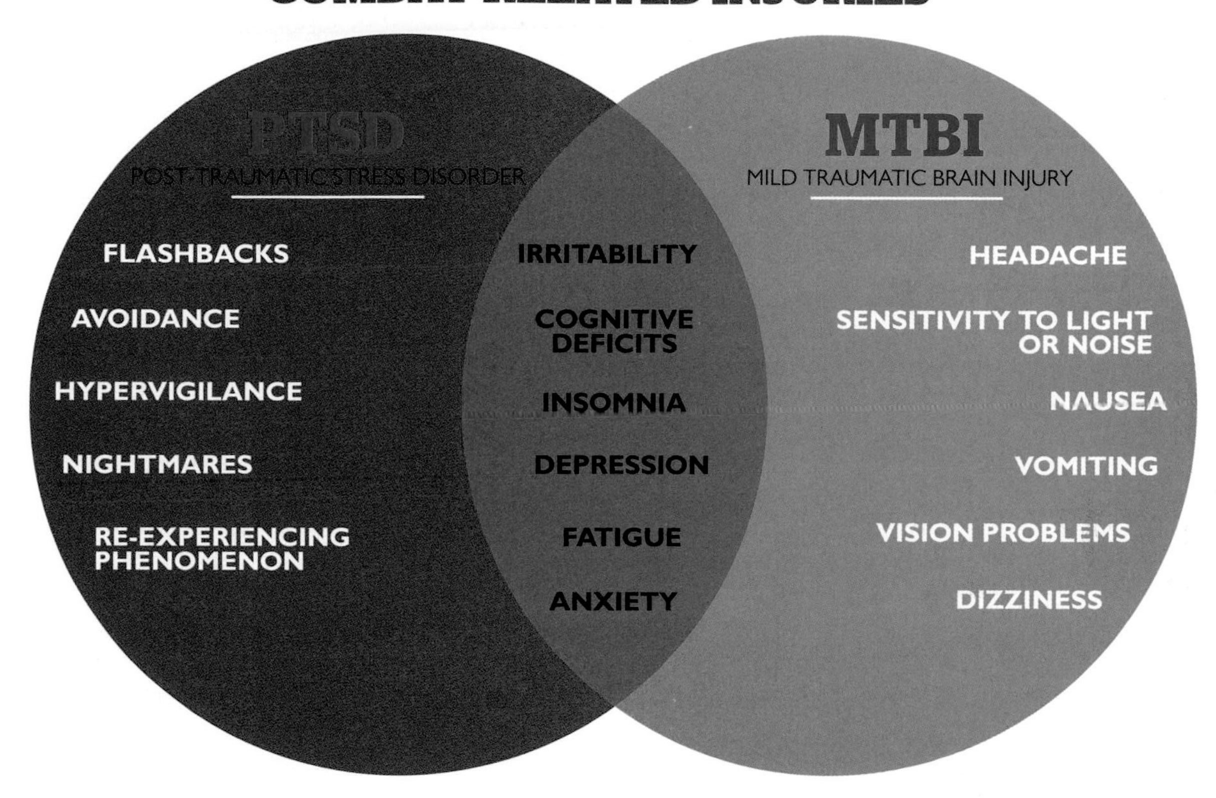

it hard to concentrate, or be hypervigilant.

The terrorist attacks of 9/11 created a grim laboratory for studying PTSD. In the two months following the attacks, 11.2 percent of New York City residents and 4 percent of all U.S. residents reported symptoms of PTSD. Unsurprisingly, the closer to the destruction, the more likely people were to have the disorder.

Combat veterans are at risk for post-traumatic stress disorder (PTSD).

Twenty percent of those living below Canal Street (near the World Trade Center) described PTSD symptoms. More than 6 percent of first responders at Ground Zero also reported PTSD symptoms.

However, the 9/11 studies also contained hope. After six months, none of the first responders had signs of PTSD any more. Symptoms of the disorder also waned in the population in general. People are resilient. About half of us experience some kind of traumatic event in our lifetimes. Only one in ten women and one in twenty men go on to develop PTSD.

» Obsessive-Compulsive Disorder

Have you ever been driving to work and wondered, Did I lock the door? Did you ever idly think about what your neighbor might look like naked? Passing a fire extinguisher in a hallway, did you ever have a fleeting impulse

People with obsessive-compulsive disorder (OCD) may be obsessed with symmetry.

to break that "In Case of Fire" glass barrier?

All of us have odd, stray thoughts and impulses at times. Most of us dismiss them as the ephemera they are. People with obsessive-compulsive disorder don't have that luxury. Their persistent thoughts return over and over again, dominating their minds and unable to be dismissed. Common obsessions among those with OCD are doubting thoughts (Did I lock the door? Did I turn off the stove?); thoughts about contamination (Are there germs on that phone? That desk? That man's hand?); and aggressive thoughts or images (What if I break that glass? What if I hurt that baby?).

The obsessive thoughts then lead to the second part of the disorder, compulsive behavior. In order to relieve the anxiety caused by the thoughts, the person feels compelled to perform repetitive or excessive actions. A woman obsessed with contamination might wash her hands over and over or feel compelled to wash them a set number of times. A man compelled by doubting thoughts might repeatedly check door handles or stove tops. Many people with OCD follow rituals, such as getting dressed in a particular order every day. They may be obsessed with symmetry—the pencils must be aligned to the side of the desk; the magazines have to be placed two by two next to each other on the coffee table. More than an occasional quirk, for the behavior to be classified as OCD it must take up a significant part of the person's time—at least an hour a day—and must cause significant distress or dysfunction.

People with OCD have trouble convincing themselves that they won't act on their inappropriate or violent thoughts. The intense anxiety these thoughts provoke can drive them to avoid everyday situations. They'll circumvent the hallway with the fire extinguisher.

They'll back out of family gatherings where children are present. Life becomes a struggle.

The disorder can appear at a fairly young age. Writer and filmmaker Lena Dunham writes that, by the age of eight, she endured a range of fearful thoughts: "The list of things that keep me up at night," she writes in her autobiography, "includes but is not limited to: appendicitis, typhoid, leprosy, unclean meat, foods I haven't seen emerge from their packaging, foods my mother hasn't tasted first so that if we die we die together, homeless people, headaches, rape, kidnapping, milk, the subway, sleep." She was obsessed with the number eight and would have to repeat behaviors eight times. About 2 to 3 percent of people develop the disorder by their teens or early 20s. Fortunately, it gradually eases with age.

»Treatments for Anxiety

Anxiety disorders, and the thought patterns that lead up to them, are highly treatable. Often doctors will combine medications with therapy particularly tailored to the patient's circumstances. No two people are alike. Cognitive behavioral therapy (CBT), which involves targeted exercises and even homework, is particularly effective as it involves the patient in actively redirecting her own thoughts and behaviors.

To relieve generalized anxiety disorder, for example, CBT practitioners might encourage a client to monitor and record her own responses. She might be coached in progressive muscle relaxation techniques and breathing exercises. The therapist could suggest picking a "worry-free zone"—say, breakfast time or driving time. If the persistent worries crop up in the worry-free zone, they're deferred to a later time and a different place. The practice of mindfulness (see chapter 8, page 252), with its emphasis on attending to the present moment, can also help create an expectancy-free, peaceful state of mind.

HELPING BY NOT HELPING

OCD sufferers don't suffer alone. Usually their parents, partners, or other caregivers suffer right along with them. Attempting to soothe their anxious companions, and perhaps simply trying to get them to move along, caregivers more often than not will help them with their rituals. They'll check that doorknob for them, lock away the knives, or answer worried questions over and over. Research shows, however, that accommodating OCD behavior simply makes it worse.

In children, the more the parents accommodate the child's behavior, the worse the child's disorder becomes. In romantic partners, not only does the "help" make the partner worse by reinforcing the behavior, but also it weakens their personal bond overall. Partners who go along with their impaired partner's obsessions report that their relationships get worse, not better. It's better not to enable a loved one, but to support them as they address the problem themselves.

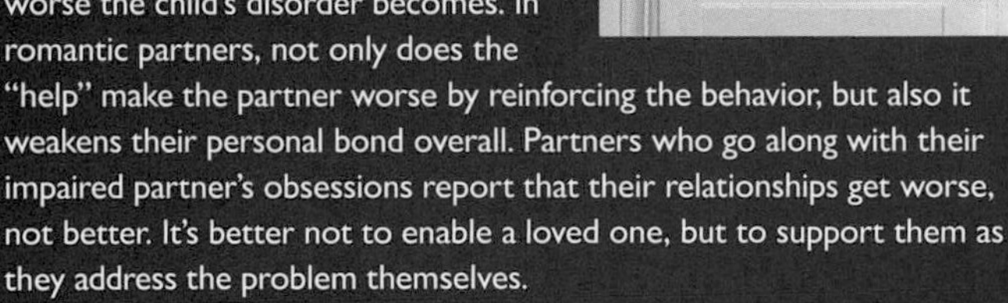

“There are strings in the human heart that had better not be vibrated.”

WRITER CHARLES DICKENS

Some anxieties, such as social anxiety, are treated with exposure therapy, in which the anxious person is gradually exposed to the fearful situation until his anxiety begins to lessen. As he goes through this, he needs to examine his own beliefs going into the dreaded situation and then evaluate how accurate those beliefs were afterward. For instance, someone who thinks he can't succeed at parties unless he rehearses all his conversations beforehand might be asked to forgo this precaution and wing it while he chats. Later, he will think about the experience and consider whether he wasn't just as successful without all his rehearsal. Relaxation therapy also helps with social anxiety. People who practice relaxing individual muscle groups (see chapter 2, page 69) can then call up this response when they feel themselves start to tighten up in social situations.

Obsessive-compulsive disorder responds to exposure therapy as well. In this approach, the therapist asks the patient to directly confront the thoughts or actions that make her so anxious and then to refrain from the accompanying compulsive behavior. Someone obsessed with germs might be asked to shake hands with a stranger and then hold back from washing her own hands. As painful as this is at first, the repeated exposure begins to retrain the brain by showing it that nothing fearsome happens afterward (see chapter 9, page 262).

This is the good news. Cognitive behavioral therapies, acceptance and commitment therapy, and other positive interventions can help most people with anxiety and other shadow syndromes. Furthermore, many treatments now won't stop at simply alleviating disabling symptoms. They will promote active well-being, helping those who are languishing to realize that they can not only survive, but also flourish.

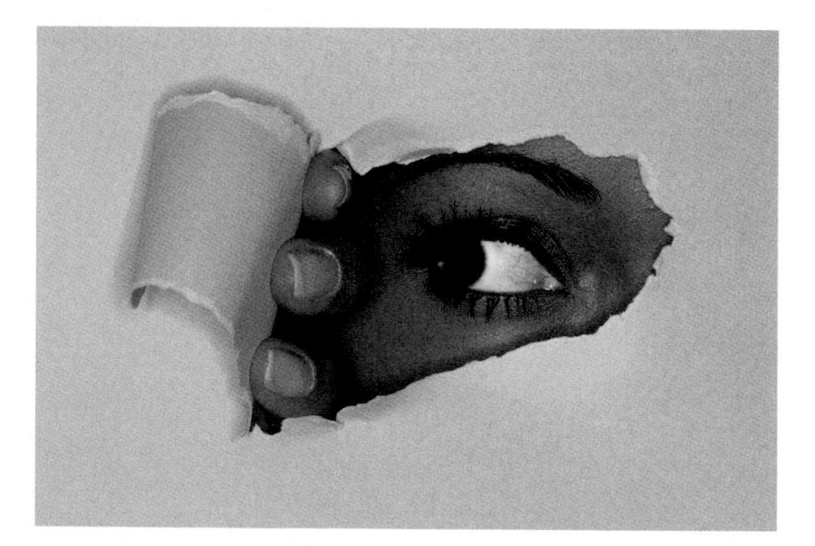

Therapy can help people break free from anxiety.

Класика майбутнього
Новий Mercedes-Benz S-Клас

» CHAPTER EIGHT «

THE ULTIMATE CONCERNS

Austrian psychiatrist Viktor Frankl, transported to concentration camps during World War II, endured tremendous suffering. During one frigid early morning work detail, as he writes in *Man's Search for Meaning*,

We stumbled on in the darkness, over big stones and through large puddles, along the one road leading from the camp. The accompanying guards kept shouting at us and driving us with the butts of their rifles . . . Hiding his mouth behind his upturned collar, the man marching next to me whispered suddenly: "If our wives could see us now! I do hope they are better off in their camps and don't know what is happening to us."

That brought thoughts of my own wife to mind. And as we stumbled on for miles, slipping on icy spots, supporting each other time and again, dragging one another up and onward, nothing was said, but we both knew: each of us was thinking of his wife . . . I understood how a man who has nothing left in this world still may know bliss, be it only for a brief moment, in the contemplation of his beloved. In a position of utter desolation, when man cannot express himself in positive action, when his only achievement may consist in enduring his sufferings in the right way—an honorable way—in such a position man can, through loving contemplation of the image he carries of his beloved, achieve fulfillment.

In the midst of horror, Frankl found meaning in love. He discovered that his life had a purpose that allowed him to transcend his grim surroundings. After he survived the death camps, Frankl went on to work as a therapist who helped others detect the meaning in their own lives.

Even as psychologists work toward identifying the troubled thought patterns that lead to mental disorders, they are also rediscovering the higher purposes that give people strength. That people strive toward goals according to their values is hardly

Strivings propel us to our goals, whether modest or grand, one step at a time.

a new idea. Entire religions and philosophies are devoted to searching for, and defining, what is good in life. Modern psychology has taken the search out of the realm of philosophy and studied it scientifically. What gives our lives meaning and purpose? How important are spiritual connections? How do all these intangible ideas affect our well-being? The results indicate that unspoken strivings, values, and transcendent experiences play a large and positive role in everyday life.

Your everyday actions on a typical day may seem inconsequential. You go to work, to the gym, to the grocery store. You take your kids to soccer practice. You email a friend. Yet your everyday actions reflect who you are and what you want out of life. They're driven by your personal strivings.

LIFE'S GPS

Strivings are meaningful objectives that people pursue through their everyday behavior. Each person has a characteristic set of strivings that reflect his own values, strengths, and commitments. They don't necessarily represent who you are right now, but rather who you want to be and what you want to accomplish. They're aimed at a goal, or a set of goals, and they're satisfied when they reach that objective.

You could think of strivings as a life GPS. (A good GPS—not the kind that sends you off a bridge.) We enter our destination and follow the directions one at a time, trusting that each move will get us closer to where we want to end up. This life map provides us with structure and order, helping us choose between the wrong and right turns.

Strivings can propel us toward grand or modest goals, and the goals themselves might be broken down into sub-goals. For instance, someone whose goal is to be more attractive might strive to get more exercise, eat less, and dress better. In his research, psychologist Robert Emmons lists one person's typical strivings:

- **Avoid letting anything upset me**
- **Work toward higher athletic capabilities**
- **Meet new people through my present friends**
- **Promote happiness and hope to others**
- **Accept others as they are**
- **Be myself and not do things to please others**

ASK YOURSELF

No matter what your age or stage in life, certain goals or purposes motivate you. These goals can be positive or negative. You might try to be a good role model, for instance, or you might try to avoid being noticed.

What do you strive for? On a piece of paper, write down eight strivings, as follows:

I typically try to ________________________________.

Your strivings can be positive or negative, broad or specific. Then, for each striving, use the following scale and choose the number that indicates how much you agree or disagree with each statement. Please describe how you have felt over the past month, including today.

Strongly Disagree		Neutral		Strongly Agree

1. **This striving is important to me and I am committed to it.**
   1 2 3 4 5
2. **In the past month I have made progress on this striving.**
   1 2 3 4 5
3. **I experience difficulty and obstacles working on this striving.**
   1 2 3 4 5
4. **I derive a sense of purpose and meaning from this striving.**
   1 2 3 4 5
5. **Pain has interfered with my accomplishing this striving.**
   1 2 3 4 5
6. **Significant others support me in this striving.**
   1 2 3 4 5
7. **This striving is not in my best interest (I am conflicted about it).**
   1 2 3 4 5
8. **I tend to experience a great deal of joy when I am successful in this striving.**
   1 2 3 4 5
9. **I tend to expend a lot of effort and energy in trying to be successful in this striving.**
   1 2 3 4 5

You can score these goals with the Striving Assessment Scale found in the back of the book (see page 290).

- **Not eat between meals to lose weight**
- **Not be a materialistic person**
- **Appear intelligent to others**
- **Always be thankful, no matter what the circumstances**
- **Reciprocate kindnesses**
- **Keep my beagles happy and healthy**
- **Do what is pleasing to God.**

Approach and avoidance (see chapter 6, page 186) apply to strivings, just as they do to other behaviors. An approach-oriented person approaches or tries to acquire the goal he strives for: "I'm going to finish that task by the end of the week." An avoidance-oriented person is trying not to do something. He wants to avoid, prevent, or get rid of the object of the striving: "I'm going to avoid procrastinating."

On the whole, an approach-oriented mind-set is usually a healthier one. Studies of combat veterans, for instance, show a clear connection between approach-directed strivings and ultimate well-being. Researchers asked veterans with and without post-traumatic stress disorder (PTSD) to describe the things they strove for in daily life. Then they asked them to keep a daily journal for 14 days to gauge their levels of well-being and self-esteem. In general, veterans with PTSD were more likely to have avoidance-oriented strivings. They tended to devote their energies to avoiding negative outcomes and to controlling their emotions. Because (like all of us) they have a finite amount of inner resources, these strivings took away from the time and energy these veterans could devote to more meaningful pursuits and more positive goals. As a result, their daily lives were marked by lower levels of self-esteem and overall well-being.

On the other hand, the veterans with PTSD who reported more positive strivings and more approach-oriented goals fared considerably better in their daily lives. Their levels of well-being were similar to those of veterans without

Highly ambitious presidents, such as Richard Nixon, may be unwilling to compromise.

"There is no simple formula for finding happiness."

PSYCHOLOGIST DANIEL GILBERT

DD
NIKE

Wanting to become closer to your family is an approach-oriented striving.

PTSD. Having PTSD doesn't mean that people are condemned to struggle. If they organize their limited time and emotional resources around positive life goals and strivings, they can do well.

»Owning Your Strivings

Where do these motivations come from? You may tell yourself that you truly want that promotion, or that you crave being closer to your family—but do you really? Striving for a goal out of guilt or because others have dictated it is a recipe for stress and failure. Psychologists recommend that you ask yourself whether you really feel ownership of your own goals and act accordingly. Be aware, though, that if you fail to get the things you truly want, you will be more strongly disappointed than if you fail to reach the goals that others have set for you

Generativity is the desire to reach out and give to others.

Psychologists today study the science of happiness.

Studies show that people who are conflicted between their strivings and their true values and interests have higher levels of depression and more psychosomatic complaints. They are less likely to achieve their goals and more likely to spend time worrying about them. Ambivalence tends to lead to stagnation. We're more successful in our

FOCUS

A STROKE OF INSIGHT

On December 10, 1996, 37-year-old brain scientist Jill Bolte Taylor underwent the kind of profound shift in consciousness that few people survive. Malformed blood vessels in her brain burst, flooding the left side of her brain and causing a major stroke. Her left hemisphere—which handles sequences, categories, a sense of time, and recognition of words and numbers—stopped working properly. Her right hemisphere—which senses the big picture, emotional states, and the present moment—took over. Along with pain, confusion, and partial paralysis, Taylor began to experience a spiritual awareness of connection to the rest of the universe.

"Deep within the absence of earthly temporality," Taylor writes in her book, *My Stroke of Insight*, "the boundaries of my earthly body dissolved and I melted into the universe . . . I'm no authority," she continues, "but I think the Buddhists would say I entered the mode of existence they call Nirvana."

Taylor's recovery from the stroke took eight years, as she re-learned how to read and control her body. But for a brain scientist, the experience was an invaluable insight into the spiritual processes of the brain.

strivings when they are truly autonomous.

LOOKING FOR MEANING

Psychiatrist Viktor Frankl suffered more than most as he sought out meaning, but even in an ordinary existence the belief that life makes sense and has a higher purpose can be a source of great strength. Researchers who study how people perceive meaning have learned that it plays a positive role in both mental and physical well-being. Those who see high levels of meaning in their existence are happier and feel they have more control in their lives. Their work is more satisfying. They are less likely than others to be anxious or depressed. They even live longer.

We see meaning in life when life makes sense to us. If our everyday lives seem understandable, if events fit into a rational and predictable pattern, we are reassured that we are living as we should. We also find meaning in transcendence, in the sense that events have a profound significance that ties into a larger picture.

Perhaps without realizing it, most of us actively seek out meaning in our day-to-day lives. When we shovel our neighbor's walk one weekend, and he shovels ours the next, it confirms our sense that people are generous. Ceremonies, too, are formal markers of meaning. A wedding, a baptism, or a graduation will

Reaffirming our values can help us face up to our own mortality, one of life's biggest challenges.

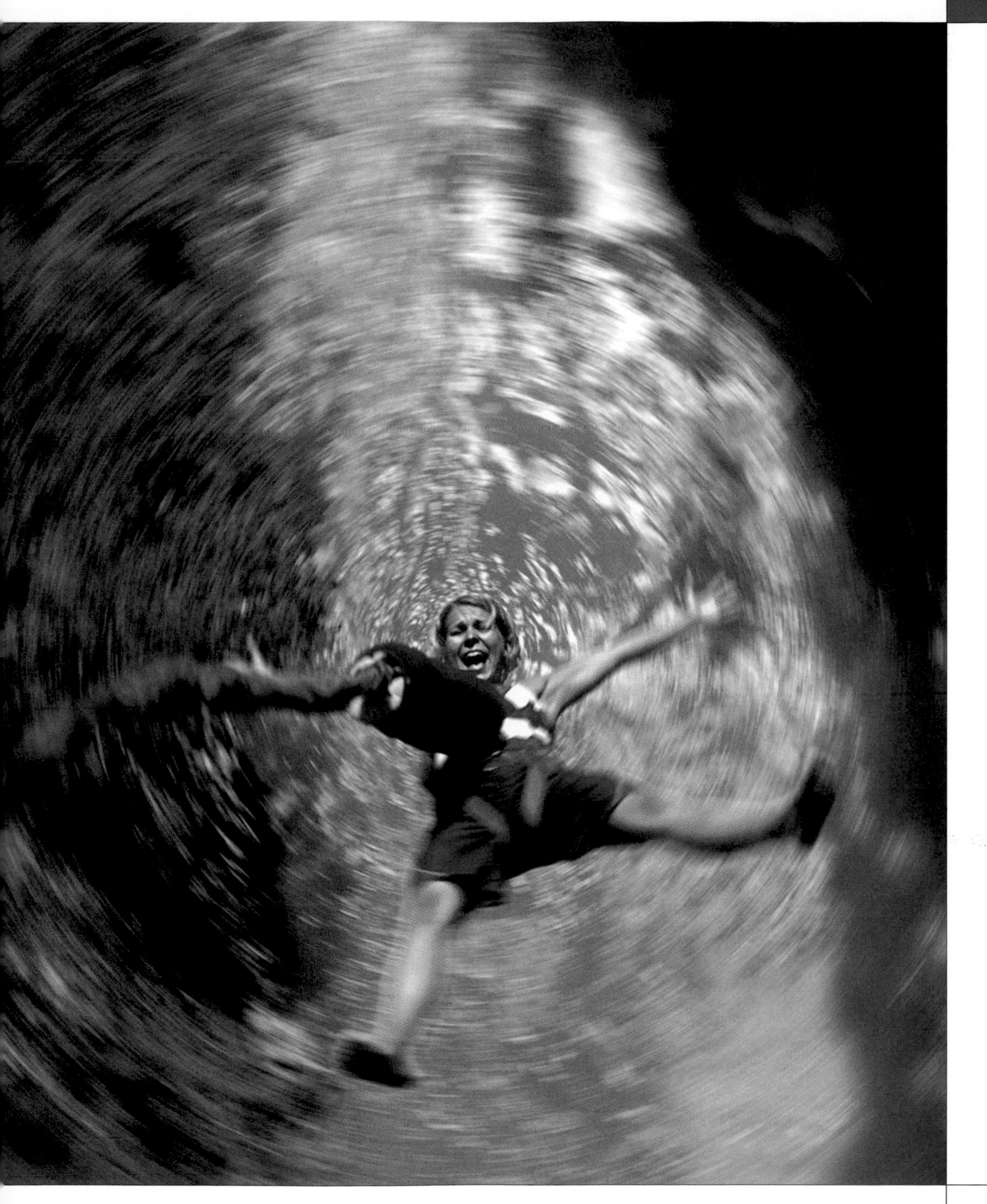

“Love is a wondrous state, deep, tender, and rewarding.”

PSYCHOLOGIST HARRY F. HARLOW

confirm our belief that life progresses in an understandable way.

At difficult times in our lives, we not only detect meaning—we create it. When events don't make sense, most of us, distressed, will work to reshape our understanding of them until our world regains its sense of order. After a loved one dies, we might console ourselves with the belief that she has gone to a better place. In the face of serious illness, we may find meaning in our own strength and in renewed appreciation of the world's beauty.

Confronted by a seemingly meaningless event, people may also compensate by reestablishing their sense of meaning in other areas of their lives. They don't always do this in a positive way. In one study, people who were told they had failed a word-association task were more likely to discriminate against another group of people. The threat to their self-esteem led them to reestablish their sense of social status by demeaning others. In another investigation, participants who were shown inconsistencies in their own lives responded by becoming more rigid in their beliefs about other, unrelated topics (capital punishment, for example).

An experiment with playing cards, conducted at Harvard in the 1940s, demonstrated just how people struggle with

People can find meaning in beauty and transcendence.

We may look for meaning by contemplating our place in the world.

Ceremonies, such as this Wodaabe beauty competition, are markers of meaning.

incongruous events. Researchers showed students pictures of playing cards and asked the students to identify them. Some cards were normal; others had their colors switched—for instance, the diamonds were black. Those who saw the color-shifted cards had one of four responses:

- **Dominance:** They simply didn't perceive the new color, and saw the black diamonds as red.
- **Recognition:** They saw the new color and adjusted their thinking to recognize that diamonds can be black as well as red.
- **Compromise:** They saw a blend of true and false information. A red six of spades might be seen as purple.
- **Disruption:** They were simply unable to process what they were seeing and couldn't respond. One anxious student commented: "I can't make the suit out whatever it is. It didn't even look like a card that time. I don't know what color it is now or whether it's a spade or a heart. I'm not even sure what a spade looks like. My God!"

In contrast to the anxiety brought on by the disruptive response, recognition represented successful meaning-making, a flexible response to a new environment. We can apply that to all kinds of disturbing or seemingly meaningless events in our lives by stepping back and accepting those events as part of a new worldview.

When times get tough, we create meaning.

»Higher Purpose

Meaning helps you make sense of your life. Purpose gives your life direction.

Your sense of your own purpose is central to your identity. It can be as basic as the idea that you must give back to your community. Creating art or caring for your family are purposes. Marines find purpose in God, country, and corps.

A sense of purpose isn't a goal in itself, an aim that can be achieved and then shelved. But it can give everyday life an invisible organizing principle, a framework for your behavior. Purpose shapes your many goals and decisions. It helps you decide how to allocate your time and resources. It can bolster your endurance in tough situations. Viktor Frankl quoted Nietzsche about purpose: "He who has a why to live can bear almost any how."

One of the most interesting aspects of purpose is that people with purpose seem to live longer. Large-scale studies over many years show that participants with a strong sense of purpose had much lower mortality rates than others. Volunteers, people who provide social support, pet owners, and those who attend religious services have longer life spans than their non-purposeful counterparts.

Purpose can prevent bad habits. A study of people with cocaine dependence showed that a sense of purpose in life predicted their ability to avoid relapse. People with purpose also seem to stay cooler under pressure. For instance, many studies have shown that individuals who are surrounded by others of different racial or ethnic groups feel increased stress. In one experiment, researchers asked participants to ride a train through an ethnically diverse area of Chicago. Those with higher levels of purpose didn't feel the increase in anxiety that their lower-level comrades experienced. Furthermore, participants who were asked to write about their sense of purpose in life for ten minutes

FINDING MEANING

Each one of us is living his own autobiography—an evolving story with significant chapters, key scenes, turning points, main characters, and lessons. We organize our experiences into stories that explain who we are, how we got here, and where we might be going. These personal narratives tell us much about how we see ourselves and larger themes in our existence. Try this little writing exercise to see where you might find growth and meaning in your own life.

Reflect on your life, beginning in childhood, and identify two areas in which growth has occurred—times when you gained insight or transformed yourself. Give specific examples with as many details as possible. Reread these incidents and identify any common themes. Ask yourself how the growth aligned with your values. Then, imagine a future scenario in which a similar moment of growth occurs. What are you doing in this scene? What did it feel like?

Don't overdo the navel-gazing. If you take apart each incident in your life, you can drain the meaning from it. Build from an assumption that life in general will make sense, and you'll do well.

Soldiers find purpose in serving their country and their units. Purpose shapes our goals.

before boarding the train also reported feeling impervious to the stress.

Purpose even seems to ward off the devastating effects of Alzheimer's disease. For example, 246 elderly participants in the Rush Memory and Aging Project took a survey gauging their sense of purpose. (They rated their agreement with statements such as "I live life one day at a time and don't really think about the future" or "Some people wander aimlessly through life, but I am not one of them.") Researchers assessed their cognitive functions over time and, after death, examined their brains for the characteristic plaques and tangles of Alzheimer's disease. They found a direct correlation between a subject's sense of purpose and better mental functioning, regardless of the actual level of damage in the brain. In other words, people with purpose seemed to have a greater cognitive reserve that allowed them to thrive in spite of the presence of disease. Discovering a sense of purpose and reflecting upon it may help any individual cope with the ups and downs of life.

A sense of purpose can be an organizing life principle.

THE SPIRITUAL SELF

In 1901 and 1902, psychologist and philosopher William James delivered a series of lectures later published as *The Varieties of Religious Experience*. In them, he considered how people throughout history have found different values in religion, which he defined as "the feelings, acts, and experiences of individual men in their solitude, so far as they apprehend themselves to stand in relation to whatever they may consider the divine." By the late 20th century psychologists began to pick up where James left off and again turned their attention to the role of religion and spirituality in people's lives. They are recognizing that, for many, spiritual beliefs give their lives purpose and meaning.

Religion and spirituality overlap: Religion typically includes a belief in a reality beyond ordinary existence, codified to some extent in moral laws, traditions, and communities. Spirituality can be defined as an inner sense of belonging, connectedness, and openness to the infinite—"a search for the sacred," in the words of psychologist Kenneth Pargament. This search may permeate everyday experiences, goals, roles, and responsibilities. Religious people are typically spiritual, but spiritual people need not be religious.

»The Role of Religion

Researchers have turned their attention back to religion in part because it seems to have a significant, positive effect on a person's well-being. Strictly from an evolutionary perspective, we can see why this might be so. Religious rules and religious communities meet many fundamental human needs. Religious groups build friendly alliances against outsiders; they band together to help the sick and elderly. Religious strictures, such as hand washing or healing rituals, can promote health. Prescribed social roles, from priests or shamans on down, maintain social order.

Today, religious and spiritual people are physically healthier than others. Religiously active folks have fewer hospital admissions, stronger immune systems, and lower levels of stress hormones. They also have strikingly lower mortality rates. A large-scale survey that followed more than 20,000 Americans over eight years discovered that

A sense of purpose helped activist Malala Yousafzai through difficult times.

"I believe in one thing—that only a life lived for others is a life worth living."

PHYSICIST ALBERT EINSTEIN

those who did not attend religious services were 1.87 times more likely to have died than those who attended frequently. Religious attendance, in other words, might add eight years to the life span. Other studies, from Israel to California, have echoed these findings.

Of course, religious attendees might have other habits that also contribute to longevity, and so they do—to an extent. Religiously active people drink and smoke less than others. Some religions require a healthy diet, such as vegetarianism. Religious communities often encourage active, physical participation. Furthermore, religion by its very nature is social, and social ties also contribute to health. Stable marriages and circles of friends keep people engaged and protected when ill health strikes.

However, these factors do not explain all of the difference in mortality between religious and nonreligious folks. The rest seems attributable to less tangible benefits—possibly the sense of meaning and hope that religious belief can impart. Spiritual beliefs help people cope with change, including the shattering experience of a loved one's death. Spiritual goals inspire people to persevere. Studies have repeatedly shown that individuals with spiritual strivings are happier than others, reporting greater life satisfaction and better marriages. People find spiritual goals to be simply more rewarding than material ones.

Spiritual beliefs are most beneficial when they are broad and flexible. People who see God as strict, angry, and punitive suffer more stress than others. These narrow perceptions can also lead to bigotry. A study of 11,000 Europeans found that those who believed "there is only one true religion" were more prejudiced than others against ethnic minorities. Views of God as purely protective and loving can also lead to cognitive dissonance when the believer is confronted with the problem of evil and suffering in the world.

» Spiritual Transformations

One of the most dramatic events in a person's life story is a religious conversion. A spiritual transformation is at the center of many famous religious episodes: Paul's vision of Jesus on the road to Damascus, for instance, or Siddhartha Gautama's enlightenment under the Bodhi tree.

A religious epiphany does not change a person's basic

Spirituality and faith can add meaning to life.

FOOD & RELIGION

Religion and food both play important roles in bringing communities together, and it naturally follows that a person's religious beliefs have an impact on their dietary habits. Many Eastern religions, like Jainism and most schools of Hinduism, mandate a vegetarian or vegan diet. The emphasis in these religious practices is to reduce suffering and violence—taking away an animal's life for your own sustenance violates this principle.

While other religions may not ban all meat, some forbid certain kinds of food, or stipulate that food must be prepared in a certain way. Practitioners of Islam and Judaism cannot eat pork, and Rastafarians avoid red meat. Many religions encourage moderation in indulgences like alcohol and desserts.

Subscribing to the set of beliefs set out by particular religion influences the way a person interacts with the world and everything in it.

personality traits. It does, however, alter her goals and strivings, as well as her fundamental life narrative. A recently converted person may change her career choice, say, from salesperson to teacher, or include new daily goals of prayer or reaching out to others. Religious experience can profoundly affect a person's sense of life's meaning.

People who have unhappy, traumatic, or stressful childhoods are more likely to experience a conversion. Those transformations often happen during adolescence, a time of stress, change, and identity search. Those who convert often report feeling inadequate or unable to form stable attachments prior to the conversion. Afterward, they feel an increased sense of purpose and competency.

»Meditation and Mindfulness

One of the oldest spiritual practices, meditation, has received renewed attention in the age of brain scans and cognitive therapy. Meditation is a mental practice in which a person directs his attention either inward or outward in a nonjudgmental way, achieving a state of relaxation and peace. Meditators leave behind ruminative, worrisome thoughts in favor of a larger spiritual awareness.

Meditation is identified with Eastern religious practices. Hindus, Taoists, and Buddhists meditate in order to transcend the self and reach enlightenment or unity with a larger reality. Western religions have a long history of meditation as well, sometimes under the names of contemplation or prayer. In the 14th century, St. Gregory of Sinai wrote, "Our aim in the life of prayer is to bring to light this divine presence within us." He suggested that the devout "sit down alone and in silence. Lower your head, shut your eyes, breathe out gently, and imagine yourself looking into your own heart . . . As you breathe out, say 'Lord Jesus Christ, have mercy on me.' . . . Try to put all other thoughts aside."

Today, meditation takes many forms, some religious, others secular. Mindfulness meditation is one of the most popular, and most studied, techniques. We've all had the experience of operating on autopilot—driving to the store without being aware of the road, eating without tasting the meal, meeting people without registering their faces. Mindfulness aims to restore our awareness of life. It means being consciously present, maintaining a moment-to-moment awareness of your surroundings, your sensations, and your thoughts and feelings, without passing judgment on them.

»Loving-Kindness

Loving-kindness meditation, another contemplative technique, is intended to build compassion toward ourselves and others. Compassion literally means "to suffer with." A compassionate person is not only sensitive to pain, but also wants to alleviate it. Self-compassion, then, means treating ourselves with the same kindness that we show others. Practicing self-compassion, we pay attention to our emotions and find ways to alleviate our own pain. Studies of brain function using fMRIs show that the same areas of the brain (the left temporal lobe and the insula) are active when people feel empathy toward others and when they reassure themselves.

Self-compassion is strongly linked to happiness and optimism. However, self-compassionate people are not Pollyannas. They don't have blindly optimistic worldviews. People who treat themselves kindly are more able than others to hold negative emotions in their awareness without either denying them or suffering excessively from them. Research shows that self-compassion is linked to reflective wisdom—

Mindfulness and meditation promote an awareness of your surroundings.

BEING MINDFUL

Being more mindful of the world around you can be good for your health. These simple exercises will help you use all of your senses.

» **Full body assessment.** From your head to your toes, focus on each part of your body and the sensations that accompany them. Take the time to appreciate and accept your body in its current form.

» **Take a deep breath.** Take a moment to feel, hear, and see your breathing patterns. Pay close attention to how your breath reflects your current emotional state.

» **Notice the ordinary.** In times where you would usually focus on your destination or tasks to be done, stop and take in the stimuli that would ordinarily fall into the background.

» **Focus on food.** Take a single piece of food and, using each of your senses, observe it in great detail. Notice the weight when you pick it up, how it feels in your hand, the texture and smell of it, and the flavor on your tongue.

» **Separate your thoughts from your self.** The emotions and thoughts that run through your head during the day don't have to define who you are.

» **Walk it off.** You don't always have to be still to meditate. Take a short walk, across the room or down a path, and focus on the way your body moves and supports itself.

self-awareness, insight, and the capacity to see life as it really is. Mindfulness is a component of self-compassion, because it involves being aware of our own emotions, including our negative feelings, in a balanced way, without exaggeration or drama.

You can build self-compassion without meditating. For instance, writing a self-compassionate letter to yourself every day for a week has been shown to increase happiness for as long as six months. However, loving-kindness meditation is also effective. In this practice, the meditator sits with eyes closed and repeats phrases of good intention ("May I be healthy," "May I be peaceful") and then directs those phrases toward others as well.

» The Benefits of Meditation

Studies of meditation have repeatedly shown that the practice relieves stress, improves physical health, and increases happiness and life satisfaction. Research on heart patients, for example, found that those who practiced relatively simple relaxation techniques such as slowing down, smiling at others, and meditating on their breathing had a 50 percent reduction in repeat heart attacks. In nursing homes, daily meditation improved mortality rates.

Meditation, at least as practiced by a master, physically affects the brain. A study of Tibetan Buddhist meditators measured blood flow in their brains via PET scans. As they meditated, their parietal lobes—the area of the brain that locates us in space and time—became much less active. The meditators were transcending the usual mental limits that define the body as they gained a larger spiritual awareness. This sense of openness to the infinite is central to spirituality across all faiths.

You don't have to be a Buddhist master to find meaning, purpose, and even transcendence in your daily life. Most of us search for meaning in small ways as we attempt to make sense of the seeming randomness of life. Many of us touch the spiritual not only in religion, but also in music, in nature, or in love. It turns out that this is a deeply healthy and positive practice that complements our quest to build our own abilities, values, and strengths.

» CHAPTER NINE «

BUILDING A BETTER SELF

In 2011, a couple of researchers at Stanford University reported a remarkable finding: Asking at-risk college students to read a short survey, write an essay, and deliver a speech raised their grades, halved the achievement gap between them and other students, and even improved their health—three years later.

The two researchers had conducted what is known as an "intervention": not the kind that involves family members persuading a loved one to enter treatment, but a brief, unthreatening exercise. They knew that some marginalized student populations, such as African Americans, had less success than others in college. For the intervention, they asked African-American college students to read a survey stating that most students felt out of place in college at first, but grew more confident over time. Then the students were asked to write an essay describing how their own experiences mirrored those in the survey. Finally, the students turned their essay into a speech, videotaped so that it could inspire others.

African-American students who completed the exercise freshman year saw their GPAs rise compared to those who didn't participate. The usual gap between their grades and those of European-American students was reduced by 79 percent their senior year. By that time, they also felt happier and healthier than their counterparts. They didn't know that the intervention had caused this—they just knew they were succeeding.

Psychology has come a long way from the days of classical Freudian psychoanalysis. Those years of sessions on the couch, probing into repressed sexual desires and parsing the hidden meanings in dreams, have given way to shorter-term techniques targeting specific problem behaviors. Simple interventions

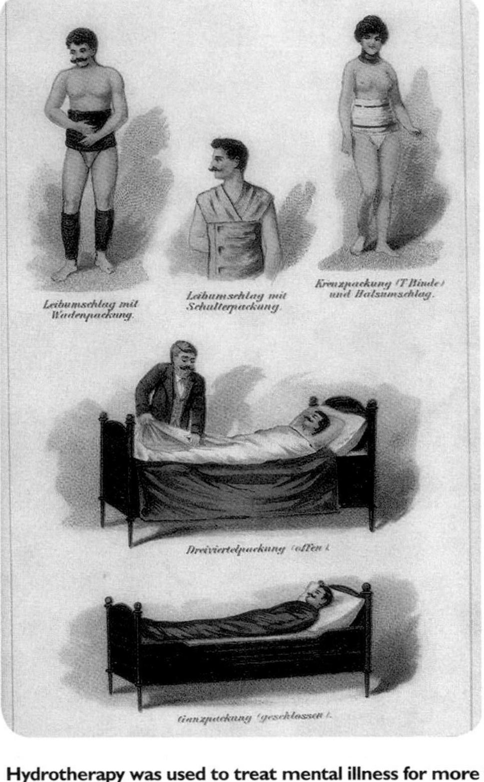

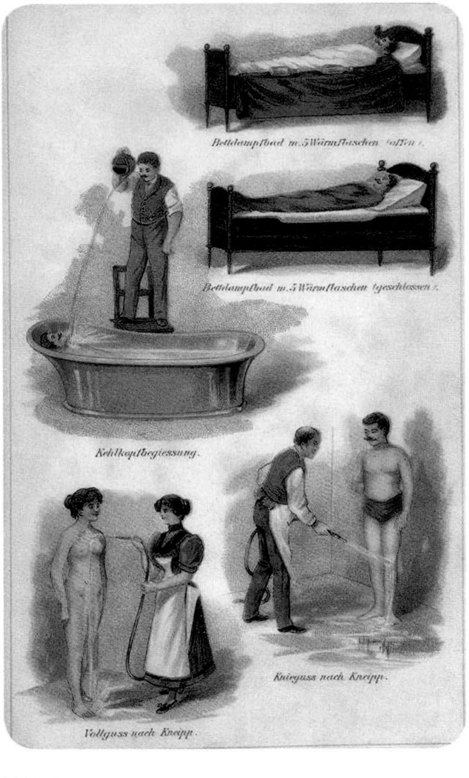

Hydrotherapy was used to treat mental illness for more than 200 years.

such as the one at Stanford have proven surprisingly effective for everyday issues. Drugs designed to fight depression or psychosis are now part of the standard psychiatric arsenal.

Psychologists are also turning their attention to everyday problems, such as the under-achievement problem addressed in the Stanford experiment. With new research in hand, they are looking at ways that the average person can live a healthier, stronger, more positive life.

TRADITIONAL THERAPIES

Patients who long for the good old days of the talking cure should realize that the road to modern therapies was a rough one that left more than a few casualties scattered along the

"Our breathing is the link between our body and our mind."

ZEN BUDDHIST MONK THICH NHAT HANH

way. Beginning in the Enlightenment and continuing well into the 20th century, well-meaning physicians employed a variety of dizzying, painful, and dangerous techniques to treat mental disorders. Usually, doctors were trying to be both scientific and humane. For most of its history, mental illness has had no effective treatment, its sufferers confined to restraints and brutal asylums. With the advent of modern neurology, many doctors grew to believe mental disorders could be treated as strictly physical ailments, curable by shocking the body or even slicing into the brain.

In the 1700s, for instance, eminent American physician Benjamin Rush recommended treating "brain congestion" by strapping the patient into a swinging chair and spinning him around. Hydrotherapy, in the form of cold baths, steam baths, or swaddling in wet linens, persisted as a treatment for agitated patients until the advent of psychiatric drugs in the 1950s. Twentieth-century treatments also included induced insulin comas and, most notoriously, lobotomies. Pioneered by Portuguese neurosurgeon Egas Moniz in 1935, this crude surgery on the frontal lobes was practiced well into the 1960s.

» Open to Suggestion

At the same time that neurologists were shocking or cutting their patients into new psychological states, many psychiatrists were employing a gentler treatment: introspection and psychoanalysis. Believing that psychological problems were functional, not physical—the result of repressed impulses, childhood experiences, and hidden interior conflicts—they concentrated on treating patients

THE MALARIA CURE

In the early 20th century, between 5 and 20 percent of men admitted to mental hospitals were given the diagnosis of "general paresis of the insane" (GPI). Their symptoms, combining psychosis and paralysis, eventually led to a gruesome death. Austrian physician Julius Wagner-Jauregg, having observed that psychotic patients sometimes improved following a high fever, began to treat GPI patients by injecting them with the blood of World War I soldiers suffering from tertiary malaria. Some of the feverish GPI patients, whose consent was not sought, died; some improved and then relapsed; others recovered their physical and mental health (following quinine treatment to cure the malaria). Wagner-Jauregg's malaria treatment caught on rapidly. The physician was given the Nobel Prize in 1927, becoming the first psychiatrist to win the award.

Why did this drastic regime work? Even as Wagner-Jauregg began his research, doctors were recognizing that GPI was a consequence of neurosyphilis, which in its later stages has both physical and mental symptoms. High malarial fevers can kill the spirochete that causes the illness. In the 1940s, with the introduction of penicillin, the fever treatment became unnecessary.

Medically, the treatment had some validity and was a forerunner of modern pharmaceutical approaches to psychiatry. Ethically, injecting malaria into unconsenting, paralyzed patients was unjustifiable. Wagner-Jauregg's methods, and his later support of the Nazi Party and eugenics, have darkened his name in the annals of psychiatry.

through discussion and interpretation. Today, traditional talk therapy has evolved into psychodynamic therapy. These modern talk therapies are a far cry from traditional Freudian analysis and have considerably more scientific backing. Even so, many people continue to believe in the efficacy of two old-style therapies: the retrieval of repressed memories and the interpretation of dreams. As fascinating as these might be to the person involved, there is little evidence that these methods work as therapy and some evidence that they can lead you down the garden path.

The brain is so suggestible that our own memories are highly fallible.

Repression, in Freud's view, was a common defense mechanism that blocked painful emotions or memories from consciousness. Sometimes these repressed memories or urges would surface in a disguised form in dreams. Most people, including some therapists, continue to believe that we bury threatening memories in our unconscious and that we will be healthier if we retrieve and acknowledge these upsetting incidents. Science doesn't back this up. Research shows that traumatic incidents, such as sexual abuse or witnessing a crime, are more likely to be sharply burned into the memory than forgotten. Sometimes an incident may be partially forgotten or deliberately ignored, only to be spontaneously recovered, perhaps triggered by a cue in the environment. But whether memories can truly be repressed and brought back only in therapy or under hypnosis is controversial.

We now know that memory is fallible and open to suggestion. One typical study, for instance, asked students simply to imagine a childhood event, such as breaking a window with their hand. Later, one-quarter of

SMASHED HIT

the students believed that this event had actually happened. In another experiment, memory researcher Elizabeth Loftus showed two groups of people a film of a traffic accident and asked one group "How fast were the cars going when they smashed into each other?" and the other "How fast were the cars going when they hit each other?" A week later, she asked each group if they had seen broken glass in the film. The people who heard the word "smashed" were more than twice a likely to say they had seen broken glass, when in fact there was none.

If a well-meaning therapist, perhaps believing that a client's behavior indicates childhood sexual abuse, probes the patient's memory with questions such as "Are you sure you weren't abused? Try to visualize the scene," it is all too possible that the patient will eventually picture just such a scenario. This doesn't mean that childhood abuse does not happen—far from it—but it's rarely forgotten. Truly repressed memories, extracted by questioning, are rare at best.

Dream interpretation, another staple of classic psychoanalysis, also fails the scientific test. Although Freud believed that a dream's latent content, its hidden meaning, revealed psychological truths, there's not much evidence that dream images are symbols for forbidden thoughts, or that dream analysis can lead to helpful insights or relieve psychological problems such as anxiety or depression. Like other avenues of analysis, dream interpretation can in fact instill false memories. In one study, for instance, many students who were falsely told that their dreams indicated that they were bullied in early childhood, or that they had gotten lost in a public place, later reported memories of these

"This is getting us nowhere!"

"You largely *choose* to disturb yourself about the unpleasant events in your life . . . You mainly feel the way you think."

PSYCHOLOGIST ALBERT ELLIS

events. Some recollections were quite specific: "My parents and I went shopping at Bellevue Square, and I ran off when they went to look at some clothes for me. I had to wait in the security office until my parents came."

» Truthiness

Drastic medical treatments such as lobotomies did more harm than good. Many psychoanalytic therapies don't stand up to scientific scrutiny. So why do we cling to these useless approaches?

Researchers, doctors, and patients alike share a common human tendency to see the world through the lens of reasonable assumptions. It makes sense that we would repress traumatic memories. It's logical to assume that a dream's dramatic images are meaningful. Some things just seem like no-brainers.

For example, "Scared Straight" programs that confront juvenile offenders with the frightening realities of prison life appear to be obvious ways to deter troubled youth from further crime. Begun in the 1970s, Scared Straight initiatives claimed success rates up to 90 percent. However, a large-scale analysis of these programs around the country painted a different picture: Kids who participated in Scared Straight were actually *more* likely to commit crimes later. The reason is unclear, though perhaps bringing young offenders together for these programs gave them unfortunate peer-group support.

We also continue ineffective

FOCUS

WHAT GOOD ARE DREAMS?

If dreams are not the "royal road to the unconscious," as Freud would have it, then what are they? Why do we dream?

We don't really know. Neuroscientists and sleep researchers have posited several theories. Dreams do incorporate daily events, emotions, and images, and in the process they may help the brain consolidate the day's memories. People who get a full night's sleep after learning a new task generally improve their skills. Sleep-deprived people, on the other hand, find it hard to learn new tasks compared to their rested brethren.

Dreams might stimulate and preserve neural pathways during sleep. Or, dreams may simply reflect random neural activity, a kind of mental white noise that the brain attempts to shape into meaning. Brain scans show that the brain's limbic system, related to emotions, is active during dreaming sleep, which may account for dreams' emotional content—typically negative, with most dreams containing some sense of fear or failure.

None of these theories is completely satisfying in accounting for dreams. They don't fully explain why we may dream of experiences we have never had, or why or how the brain is able to make a coherent story from random signals.

Patients and doctors will look for any small signs of improvement.

treatments because we believe evidence that confirms our preconceptions and discount evidence that doesn't. Patients want to improve; their doctors want them to improve; both parties will grasp at signs of improvement and discount evidence to the contrary. The illusion of control bolsters this tendency. We all want to think we can control our lives, so we're invested in believing that our efforts are working.

No scientific evidence backs up dream interpretation as therapy.

Finding a cause-and-effect relationship where none exists is another common error. Just because we are going to therapy and improving at the same time doesn't automatically mean that the therapy is causing the improvement. Correlation is not causation, as they say. (Weddings and suicides both spike in June, for instance, but that doesn't mean that one causes the other.)

And of course, most of us resist change. We're used to certain kinds of therapies. We've read about them for years. We see them on TV. It's hard to turn away from the status quo.

TREATMENTS THAT WORK

To avoid these natural errors, and to promote more effective treatments, psychologists push for evidence-based practices (EBP). If you are considering any sort of therapy, you should look for three things:

- **The best research evidence.** Your therapy should be supported by scientific research, including, if possible, randomized clinical trials.
- **Clinical expertise.** Among other things, your therapist should have a scientific background, be up to date in his or her skills, understand the current research, and have a clear rationale for treatment.
- **An understanding of the patient's characteristics and preferences.** This should include his or her cultural background and values.

Patients need to be active and informed partners in their own treatment. Gone are the days of Doctor Knows Best.

» Cognitive Behavioral Therapy

The most widely researched and scientifically supported therapy today is cognitive behavioral therapy (CBT). This practical, goal-oriented treatment aims to alter maladaptive thinking (that's the cognitive part) and self-destructive behavior (that's the behavioral part). In CBT, people identify problematic thought patterns, sometimes called "thinking mistakes," and work on replacing them with more helpful thoughts. They may also practice recognizing and replacing their unproductive behavior with more helpful actions.

Thinking mistakes typically take the form of unwarranted assumptions and negative, even catastrophic expectations of life (see sidebar on page 265). Most people fall prey to this kind of thinking at times, but when it begins to dominate your moods and attitudes, it can propel you into dark emotions, bad decisions, and irrational actions. That's why CBT patients become students of their own minds. In CBT, a therapist might ask a patient to keep a daily journal of his thoughts. The client might be assigned behavioral homework. For instance, someone with obsessive-compulsive behavior will learn to recognize and

EVIDENCE-BASED PRACTICES (EBP)

Best Research Evidence
EBP
Clinical Expertise
Patient's Values

Cognitive behavioral therapists ask their clients to look closely at themselves.

I'm not locked into this position.
I can shift with the circumstances.

record his compulsive thoughts. His therapist may ask him to gradually expose himself to the feared stimulus without resorting to compulsive behavior—touching a dirty doorknob without washing his hands afterward, for instance. Studies have shown that most OCD sufferers see their symptoms fade after this sort of treatment.

Cognitive behavioral therapy works well with a range of problems, including anxieties and phobias, depression, post-traumatic stress disorder, panic attacks, and eating disorders. The method sounds simple, but it has genuine physical effects. After months of treatment, CBT patients' brain scans show positive changes, a return to more typical processing.

» Flexibility

Other, newer approaches to therapy blend some of the methods of cognitive and behavioral therapy with recent insights about emotional regulation, personality, values, and mindfulness. Acceptance and commitment therapy (ACT), for instance, builds psychological flexibility.

Flexibility is both physically and mentally healthy.

ASK YOURSELF

ARE YOU THINKING STRAIGHT?

Cognitive therapy helps a patient recognize his own problematic thinking and address it. Do you make any of these mental mistakes?

Black/white thinking
You view a situation or person as all good or all bad, without noticing any points in between.

Predicting the worst
You predict the future negatively without considering other, more likely outcomes.

Missing the positive
You focus on the negatives and fail to recognize your positive experiences and qualities.

Treating feelings as facts
You think something must be true because you feel it so strongly, ignoring evidence to the contrary.

Jumping to conclusions
You decide that things are bad without any definite evidence.

Mind reading
You assume that you know what others are thinking without asking.

Fortune telling
You predict things will turn out badly.

Assuming control
You assume that you can control how others behave in situations where you really don't have any control.

Expecting perfection
You believe that you (or others) should be perfect in the things that you (or others) say or do.

If these negative thought patterns sound familiar, challenge them by asking yourself:

Do I know for sure that X is going to happen?
Does X always lead to Y?
What is the probability of X happening?
What is my evidence for X? What is my evidence against X?
Is there another way of looking at the situation?
What would I say to a friend who had a similar thought in a similar situation?
How will I feel about this tomorrow? In one month? In six months' time?

TAKE A FLEXIBLE APPROACH

Incorporating aspects of acceptance and commitment therapy (ACT) into your life can help you become a better problem-solver. Try this exercise for coming up with creative solutions to life's obstacles.

Identify the problem.
What feelings or tasks do you struggle with regularly? Think about a specific issue in your life and identify the ways you've tried to deal with it in the past.

Reflect on past solutions.
Do you try and fail to control your emotions, avoid the problem entirely, or take some other approach? How well have these solutions worked?

Consider alternatives.
Think of some new ways to address your problem, even if they're not methods you would normally pursue. What might you suggest to a friend with a similar problem?

Make a plan.
Commit to trying new solutions that align with your personal values and goals. Don't limit yourself to a single response.

People with emotional disorders are often locked into rigid responses: Someone with social anxiety, for instance, might fear and avoid all parties. Depressed people typically bring the same dimmed reactions to any experience, no matter how stimulating. People who are more psychologically flexible, on the other hand, are able to respond to different situations with different strategies, to shift their mind-sets when their fixed attitudes get in the way, and to adapt their behavior to their goals.

ACT aims to promote flexibility through:

- **Acceptance** Learning to accept thoughts and feelings without trying to change them.
- **Contact with the present moment** Mindful awareness of and direct contact with the present moment.
- **Defusion** "Defusing" or detaching thoughts from reality, seeing thoughts as just thoughts. Rather than thinking, "I am a bad person," thinking, "I am having the thought that I am a bad person."
- **Self as context** Observing yourself from different perspectives.
- **Values** Identifying personally relevant values.
- **Committed action** Committing to behaviors that are in line with one's values.

Accepting, not suppressing, negative feelings is a key part of ACT. Most of us are conditioned to hide or suppress anger, but that's not always the best strategy. For instance, in an experiment where subjects were asked to play the part of a landlord trying to get a tenant to pay overdue rent, the people who enlisted their angry emotions were more successful. Anger can be appropriate when you're bargaining, or when you need to confront injustice or oppression.

»Many Troubles, One Treatment

If it seems as though there's a confusing array of therapy options—there is. Some psychologists recommend taking a unified approach to emotional disorders such as anxiety or depression.

Social anxiety, phobias, and depression (unipolar depression, as opposed to bipolar) can be expressed in very different ways, but they may all be members of the same dysfunctional family of disorders. People with one such problem—one kind of anxiety

Cognitive and behavioral therapy can help patients avoid catastrophic thinking.

issue, for instance—often have another disorder as well, such as depression or a different form of anxiety. (Doctors call this "comorbidity.") Treatment for one disorder often helps with another one; people successfully treated for panic disorders often find that their generalized anxiety or depression is lessened, as well.

A common theme of emotional regulation seems to run through these conditions, pointing the way to a common treatment. The ability to respond appropriately to emotional situations, whether positively or negatively, is an important part of mental health. A unified treatment for mood disorders would use some of the techniques of cognitive and behavioral therapy across a range of problems. A therapist might ask her client to modify his flawed way of appraising situations (including catastrophic thinking, such as "This

People with one social disorder are likely to have others.

“No doubt the imagination of patients often has an influence upon the cure of their maladies.”

BENJAMIN FRANKLIN

is going to be a disaster") and to change emotion-driven behavior. This can be as simple, yet as difficult, as asking an anxious, perfectionist patient to deliberately leave things messy and unfinished, or requiring a panicked client to mingle with a crowd and smile. Therapists may also direct their patients not to suppress threatening emotions, but to recognize them and so to recognize the problematic behavior that comes with avoidance.

» Psychodynamic Therapy

Is there still room for Freud and psychoanalysis on the crowded shelves of modern therapies? The sage of Vienna has in fact undergone a bit of rehabilitation in recent years. Many of this theories have been discredited: Repressed incestuous drives, the death instinct, ids and egos, penis envy, and other mainstays of Freudian thought have been shown to be unfounded and unscientific. And yet Freud had some profound insights into human behavior that have stood the test of time. These insights now form the core of psychodynamics, an updated version of psychoanalysis.

Psychodynamic theory recognizes that much of our mental life is unconscious, driven by emotions and behavior that we can't rationally explain. Neurology supports this, showing that our brains store related processes in separate regions. Separate pathways in our brain can result in conflicting feelings and motivations. You can need but dislike someone at the same time, or desire and yet fear success. Though Freud's view of childhood was skewed, he was right in thinking that childhood experiences can affect our personalities and behavior. The road to maturity certainly involves

Freud was right when he said childhood experiences affect adult life.

learning to control sexual and aggressive impulses.

Therapy based on psychodynamic theory focuses on helping patients gain insight into the hidden processes and competing desires that steer their problematic behaviors. For example: Julie, 57 years old, is seeing a therapist in part because she has trouble dealing with her son. She loves him and is proud of him, but finds that she lashes out at him with sarcastic or belittling comments. In therapy, Julie is examining her own relationship with her parents, who were highly critical and unsupportive. She is realizing that she learned to distrust and disbelieve praise; sarcasm became a defense against her own vulnerability. Therapy can show her that her need to protect herself is conflicting with her love for her son and desire to be a good parent.

> **Character strengths are expressed in habits and in actions, not in intent.**

No longer is there an unbridgeable gulf between brain science and psychotherapy, or indeed between any two mainstream therapies. Freud might disagree with contemporary approaches, but he would still recognize some of his core concepts within them.

»A Positive Approach

Therapy is invaluable for

handling the dark side of our lives—our illnesses, crises, and conflicts. But what about the light side? Most of us, most of the time, are doing okay. What we would really like is to do better: to be happier, more productive, more connected.

Happier people have longer, healthier, and more successful lives. Their marriages are more stable, their immune systems stronger, and their incomes higher. They're even more creative than their peers. And though the arrow points both ways—certainly wealth and health make you feel better—studies show that positive emotions alone will contribute to long-term well-being.

It makes sense, then, to consciously work to improve your own well-being, and programs exist now to do just that. Positive psychological interventions (PPIs) are easy, regular exercises that imitate the healthy thought patterns of naturally happy people. For instance, happier people tend to be more consciously grateful for the good things in their lives. So an equivalent positive intervention asks people to write a letter of gratitude to someone who helped them (see sidebar on page 272). The exercises may seem almost too simple to be effective, but studies have shown that people who take part in them end up significantly happier.

A few conditions are necessary for the activities to work best. They include:

- **Motivation.** You need to really want to feel better (not everyone does!) and you need to believe the exercise will work. You have to be ready to put some effort into it.
- **Dosage.** Just like medicine, the activities work best only in certain quantities at certain times. For instance, one study showed that performing five kind acts in one day left participants happier than spreading those same five acts throughout the week. Counting your blessings is more effective once a week than three times a week. Why? Maybe for the simple reason that we find it easier to follow a once-a-week schedule for all sorts of experiences. Or maybe repetition turns the proper dose into an overdose.
- **Autonomy.** We're more successful when we feel we're in control. People who choose to take part in PPIs end up feeling better than those who are assigned the exercises. We're also more successful when we can discover benefits for ourselves. In one study, people who heard instructions and

"O love, sweet madness! Thou who healest all our infirmities!"

PERSIAN POET RUMI

TRY IT

THE HOME HAPPINESS TOOL KIT

Positive psychological interventions are brief, cheap, and easy to do at home. Here are some you can try for yourself:

Write a gratitude letter. Write a letter to someone who did something particularly kind for you but was never properly thanked. You can deliver the letter if you like, but that's not essential.

Name three good things. Write down three good things that happened today and the causes for each one.

Savor the moment. Focus on an enjoyable everyday activity, such as talking to friends or drinking coffee, and consciously savor the experience.

Count your blessings. List people or events for which you're grateful.

Imagine your best self. Visualize your best possible self—perhaps in an ideal future, when you've achieved your goals and dreams. Write down this vision as a guide to your decisions today.

Perform kind acts. These can be as simple as taking over a household chore or feeding your pet a new treat.

testimonials from others about a gratitude exercise seemed to lose the incentive to start or complete it on their own.

- **Variety.** It's still the spice of a positive life. Those who performed varied acts of kindness each week feel better than those who perform the same ones over and over.
- **Social support.** Yes, we want to do things our own way, but once we get going, a little help from our friends can strengthen our resolve. This applies even to support via social media.

Positive interventions work only to the extent that they bolster positive emotions, thoughts, and behaviors—all of which contribute to happiness. How do they do this?

Among other things, these exercises seem to satisfy some basic human needs, including the needs for autonomy, connectedness, and competence (see chapter 6, page 183). Some neurological studies suggest that positive emotions affect the brain's neurotransmitters, specifically dopamine, which is associated with rewards, mental flexibility, and extraversion. Performing positive actions has been shown to have healthy

Everyone has top strengths that are particular to him or her.

spillover effects in other areas of life. For example, people who worked on counting their blessings ended up spending more time exercising and even slept better. PPIs seem to brighten a person's general outlook, improving his or her view of naturally occurring events.

As with any activity, PPIs need to be tailored to the individual. If you hate writing letters, then you're not likely to stick with gratitude letters or gain by them. Older people seem to benefit more from these exercises, possibly because they have the time and are more willing to persist; those from the more hedonistic Western cultures see greater improvements than those from the East.

» Using Your Strengths

Positive interventions can also take the form of building on

Positive interventions can fulfill the need for human connection.

FREE
HUGS

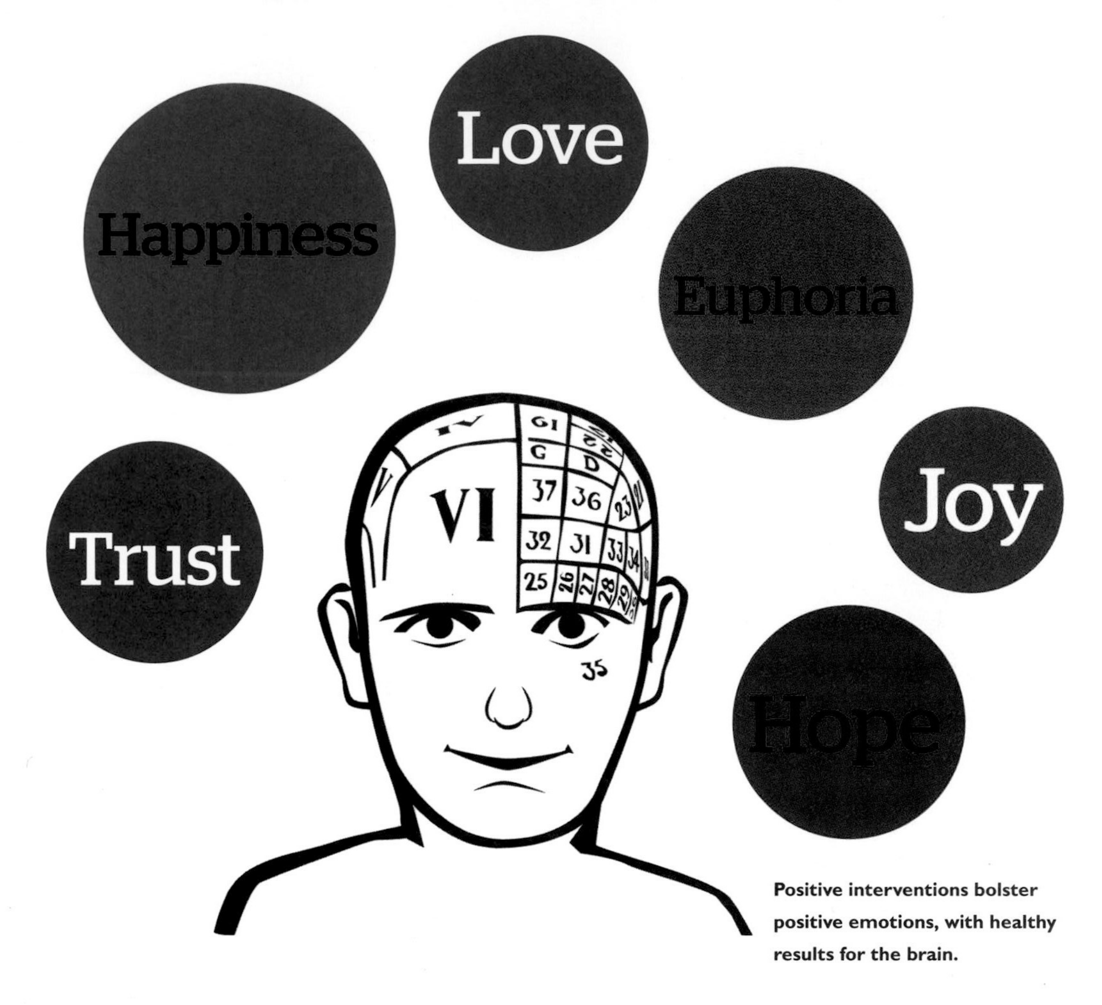

Positive interventions bolster positive emotions, with healthy results for the brain.

your character strengths. These are the qualities that give us the greatest success in relating to others and to the world around us. Everyone has top strengths that are particular to him or her—your strongest strengths, you might say.

Some researchers have codified strengths into six broad categories, each one with its own subset of adjectives. They are:

- **Wisdom and knowledge**—including creativity, curiosity, open-mindedness, love of learning, and perspective
- **Courage**—including honesty, bravery, persistence, and zest
- **Humanity**—including kindness, love, and social intelligence
- **Justice**—including fairness, leadership, and teamwork
- **Temperance**—including forgiveness, modesty, prudence, and self-regulation
- **Transcendence**—including appreciation of beauty, gratitude, hope, humor, and religiousness/spirituality

The strengths most strongly associated with personal well-being are love, gratitude, hope, zest, and curiosity. Perseverance, love, gratitude, hope, and perspective predict academic achievement among students. Effective teachers have high social intelligence, zest, and humor. Bravery and an appreciation for beauty help in recovery from illness. Perhaps surprisingly, the best indicator for an effective military leader is a high score on love. Humor predicts how much their subordinates trust them, while perspective predicts how much their leaders trust them.

Character strengths are expressed in habits and in actions, not in intent. They have varying effectiveness depending upon the situation. For instance, the most effective strengths at work are honesty, judgment, perspective, fairness, and perseverance. The least useful in the workplace: religiousness/spirituality, appreciation of beauty, love, bravery, and modesty.

Strengths will lie dormant if they're not actively harnessed. If you're a creative person, you can build upon that strength only if you can exercise it at work or at home. Strengths are also situational: That creative strength will come into play when you're designing a newsletter, while leadership comes to the fore when you're supervising the newsletter's writers. To boost your well-being, you can try the simple but effective activity of finding one new way to use your best strengths each day.

» Wise Interventions

In 2008, two groups of California voters filled out similar surveys. In one, they were asked, "How important is it to you to be a voter in tomorrow's election?" The other group was asked, "How important is it to you to vote in tomorrow's election?"

On election day, the people

Take this survey to find out how much you use your strengths in daily life. Answer all questions using a seven-point scale, ranging from 1 ("Strongly Disagree") to 7 ("Strongly Agree").

1. **I am regularly able to do what I do best.**
2. **I always play to my strengths.**
3. **I always try to use my strengths.**
4. **I achieve what I want by using my strengths.**
5. **I use my strengths every day.**
6. **I use my strengths to get what I want out of life.**
7. **My work gives me lots of opportunities to use my strengths.**
8. **My life presents me with lots of different ways to use my strengths.**
9. **Using my strengths comes naturally to me.**
10. **I find it easy to use my strengths in the things I do.**
11. **I am able to use my strengths in lots of different situations.**
12. **Most of my time is spent doing the things that I am good at doing.**
13. **Using my strengths is something I am familiar with.**
14. **I am able to use my strengths in lots of different ways.**

Scoring: Scores run from 14 to 98, with higher scores indicating greater strength use.

People respond positively to being identified as voters, which suggests that they are active and responsible citizens.

who received the noun-oriented, "be a voter" survey voted in substantially greater numbers than those who answer the verb-oriented, "to vote" question. Just a few words increased participation by 11 percent.

Small changes can have big effects, and this insight guides the science of wise interventions. Called "wise" because they target very specific processes, these interventions are brief, natural actions that focus on interrupting and changing harmful patterns in daily life. In the case of the motivated voters, the simple word change apparently affected the participants' view of their own identity. The phrase "be a voter" represented the opportunity to become a desirable kind of person, an active and responsible citizen.

> A full night's sleep after learning a new skill aids retention.

Wise interventions can succeed in a wide range of settings, from civic life, as seen in the voter survey, to education, health, and relationships. They work when they are grounded in theory and tailored to a person's specific background, motivations, and needs. In particular, the interventions need to interrupt a repeated process, a self-reinforcing cycle of behavior, so that they can have long-term effects. For instance, married couples in conflict were asked every four months to write down how a neutral third party would view their fights. Putting their problems into a dispassionate perspective helped the couples get less angry when conflicts arose again, which helped them deal with future fights, and so on. These couples found that their stresses eased and their marriages stabilized, as opposed to a control group that didn't do the exercises and whose relationships continued to go downhill.

Properly focused, these exercises can do great good. After identifying new mothers at risk of harming their children, for example, interviewers talked to the mothers until the women identified non-self-blaming and non-child-blaming reasons for their difficulties. With this perspective, the interviewers asked the mothers to brainstorm possible solutions. Compared to a control group, these empowered parents and their children fared much better, with a 4 percent incidence of abuse, compared to 25 percent in a control group.

»Affirming Your Values

Interrupting a harmful process is one form of wise intervention. Reinforcing a positive mind-set is another. All of us, consciously or not, live by a set of values, beliefs that motivate us to act in certain ways. We all, consciously or unconsciously, rank those values in a personal hierarchy. We might place conformity over independence, for example, or achievement over security. We can hold conflicting values within ourselves, and our values may conflict with those of others.

Researcher Shalom Schwartz finds that certain basic values can be found across all cultures, although their relative rankings will vary from group to group.

Couples in conflict benefit from seeing their arguments as an outside party would.

“Psychoanalysis is in essence a cure through love.”

PSYCHOANALYST SIGMUND FREUD

They are:

- **Achievement:** personal success through demonstrated competence
- **Benevolence:** preserving and enhancing the welfare of those close to you
- **Conformity:** restraint of actions, inclinations, and impulses likely to upset or harm others
- **Hedonism:** pleasure or sensuous gratification
- **Power:** social status and prestige, control, or dominance
- **Security:** safety, harmony, and stability of society, of relationships, and of self
- **Self-Direction:** independent thought and action
- **Stimulation:** excitement, novelty, and challenge
- **Tradition:** respect, commitment, and acceptance of the customs and ideas that one's culture or religion provides
- **Universalism:** appreciation, tolerance, and protection of all people and for nature.

Wise interventions help parenting skills and reinforce a positive mind-set.

In general, the rankings will skew according to gender. Men consistently value power, achievement, hedonism, stimulation, and self-direction more than women do. However, too big a difference between value sets can be bad news for a couple. Having similar values is even more predictive of a satisfying romantic relationship than having similar personalities.

Almost all parents want to pass on their values to their children. It's a hallmark of successful parenting, a way to transmit family identity from generation to generation. Studies have reached the unsurprising conclusion that when children adopt their parents' values, their family relationships are closer and smoother. But the best way to achieve this transference of values is not by imposing them from the outside. Instead, parents need to let

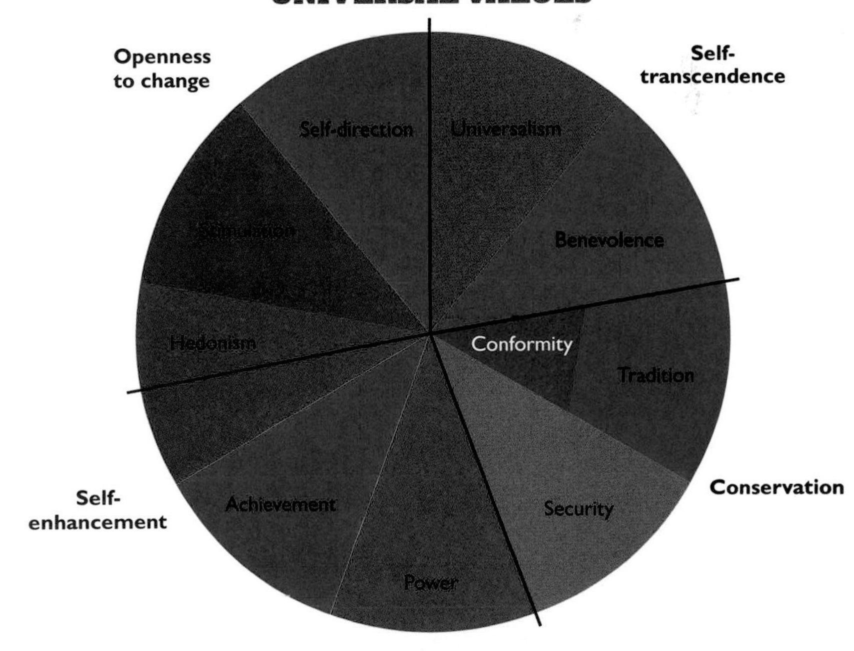

children think for themselves—to support particular values because they judge them inherently worthy.

Adolescents in general value openness and growth, but tend to believe that their parents are more conservative and less supportive of these values. However, children whose parents encourage autonomy not only find themselves more likely to agree with their parents' values in the long run, but also report a greater sense of well-being. They

> **Happier people tend to be more consciously grateful for the good things in their lives.**

agree with statements such as, "When my parents make decisions, they try to consider what I want," "My parents enable me to find my own personal way to express the principles they believe in," and "My parents try to answer seriously the questions I have regarding their principles or the behaviors important to them." Parents who support their children's choices find that, in the long run, their children will support theirs as well.

For all ages, values can be sources of strength in a frightening world. Consciously affirming our values appears to bolster their empowering effects. For instance, people who listed their top values and then wrote an essay about them had better

Children who willingly adopt their parents' values have closer family relationships.

self-regulation when faced with difficult tasks or even pain. Female college students who wrote about personally relevant values for just 10 to 15 minutes saw their grades go up in science and math courses. The biggest increases went to the women who had been mostly likely to believe that men were naturally better in physics.

Minority students, as we saw at the beginning of the chapter, are particularly vulnerable to the self-defeating effects of stereotypes when it comes to school performance. Values affirmations can help turn this around. In one intervention, at-risk African-American middle school students saw their grades soar after they wrote a series of essays on important personal values, such as family relationships or music. Two years later, the positive effects of the intervention could still be seen. Grade-point averages for all African-American students who had written about meaningful values had risen by an average of 0.24 points. For the most at-risk students, grades had risen 0.41 points. Compared to a control group, the formerly low-achieving students were much less likely to have been sent to remediation or held back a grade. The early interventions seemed to have broken a cycle in which low expectations created low performance, which then further depressed expectations in a downward spiral.

Reaffirming values even helps with one of the most challenging problems we face: acknowledging our mortality. Studies have found that people who affirm their own values are less vulnerable to death-related thoughts than those who do not. Thinking about our values may lend order to an otherwise chaotic existence, provide our own lives with meaning, and allow us to see ourselves as part of a cultural fabric that will exist after our own death, giving us a kind of immortality. Values affirmation can have a downside as well, when by reinforcing our own worldviews we narrow our perspectives. People who reaffirm their values before evaluating challenging statements (such as political views contrary to their own) are less open-minded and less able to negotiate the new information. Values, it appears, work best as a foundation, not a barrier.

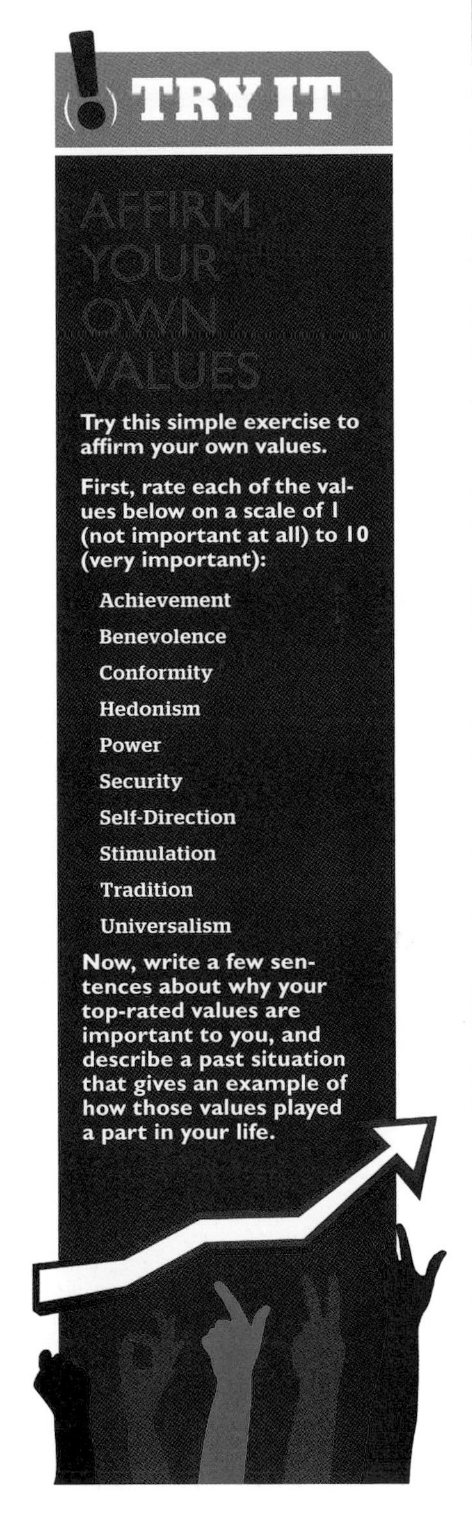

» Natural Interventions

You might not realize that you are already making positive changes in your own life simply by following some natural, pleasurable activities. We don't always need formal interventions, especially if we recognize and capitalize on the benefits of these daily actions. They include:

- **Sleep.** A good, uninterrupted night's sleep boosts mood, attention span, memory, athletic performance, and grades in school. People who sleep well have lower stress levels. Most people need eight hours, and teenagers at least nine.
- **Exercise.** Not only does it strengthen your body, it bolsters your mood. Just five minutes of exercise can improve your outlook. Regular exercise works as well as antidepressants to improve the mood of people with depressive disorders.
- **Alcohol.** You may not realize this is a natural intervention, but it's a classic. Yes, drinking can be dangerous when it's overdone. But in moderation, it's a well-known social lubricant, reducing anxiety and bolstering courage.
- **Sex.** It's not just fun in the moment: Sex can improve mood and reduce anxiety. In one study, for instance, people with social anxiety felt less anxious the day after having sex. The more intimately connected they felt to their partner and the more pleasurable the experience, the less anxious they felt.
- **Prayer.** Prayer adds meaning to many people's lives and helps them cope with illness and difficulties. Studies have shown that people who were assigned to pray for their romantic partners were more likely to forgive them.
- **Meditation.** Meditation (see chapter 8, page 250) can bring peace and relaxation, relieving stress and helping you gain a more balanced perspective on life.

Simple, pleasurable activities are psychologically healthy.

Science has made great progress in the quest to understand the human mind, but we're still on the early stages of that journey. The brain is a fantastically complex and powerful organ. We're just starting to pick apart the interplay of thoughts, emotions, desires, and behaviors that make up the mind. Even so, we've learned enough to begin to understand our own selves. Happily, with that understanding comes a sense of direction. We know enough that we can now take steps to make our lives better by overcoming our fears, building on our strengths, and connecting with others in a healthy, productive existence.

APPENDIX

PAIN AS SOCIAL GLUE

BY TODD B. KASHDAN, PH.D.,
AND ROBERT BISWAS-DIENER, PH.D.

In 1967, one of the most bizarre musical pairings in history occurred. Psychedelic rock musician Jimi Hendrix was given an opportunity to open for the Monkees. The Monkees wanted to be legitimized as serious musicians and Hendrix's band had yet to gain a following in the United States.

Imagine the cultural implosion on that first tour date, in Jacksonville, Florida. The family-friendly fans of the Monkees were taken aback by Hendrix in a neon-colored blouse wailing violently on his guitar before setting it on fire. When Hendrix asked the crowd to sing along to "Foxy Lady," they drowned him out with chants of "We want Davy!!" After offering the middle finger to the jeering crowd, Hendrix and his band left the stage and, ultimately, the tour. Instead of altering their sound to win over pop-music lovers, the three band members endured this tough experience, solidified their musical identity, and, not long after, revolutionized modern music.

It turns out that people are hardwired to connect through pain. A study conducted last year by Dr. Jim Coan and his colleagues at the University of Virginia illustrates this point. The researchers wanted to know whether physical threats to a close friend—electric shocks to the ankle in this case—led to a pattern of brain activity that was similar to when shocks were administered to strangers. When a stranger or close friend received electrical shocks to the ankle, nearly everyone responded negatively (good news—few people take joy in the suffering of others). In a new study condition, however, the participants themselves were administered mild shocks. The researchers discovered the exact same brain activity when someone received a personal shock as was activated when the shock was delivered to their friend, but not to a stranger.

From an evolutionary perspective, this makes sense. To increase the probability of survival we need to find people we can trust and people who can add their resources to our own. When we travel to a foreign country, it is helpful to have a companion who speaks the local language. When we are hiking and approach an intimidating scramble, it appears less steep when we are standing beside a close friend. Our brains treat close, reliable people in our social networks as part of the self—resources that we can depend on when in a crunch. This ability to acquire strength from the presence of other, reliable people is human evolution at its finest.

Pain, as it turns out, offers a shortcut to forming mutually beneficial relationships. A team in Australia recently investigated whether shared pain fosters social bonding. In one study, groups of strangers were randomly assigned to either painful or similar but painless tasks. Painful tasks included submerging hands in ice-cold water or holding an upright wall squat for a lengthy period of time (the painless versions involved room temperature water or simply balancing on one leg with permission to grab the wall as needed). After these opening tasks, the respective groups had to find metallic balls in a large water tank and organize them in corresponding containers at the bottom of the tank. The people who had endured the painful tasks felt a greater sense of group loyalty and showed a boost in cooperation while completing the subsequent challenge.

Sounds great if you want to locate arbitrary objects with a group of strangers, but do these types of laboratory studies generalize to daily life? They do. Shared painful experiences speed up the intimacy process. This is why people who lent a helping hand on 9/11 or who rescued survivors during Hurricane Katrina sometimes feel a lifelong bond. In a sense, these new friends are added to our very sense of self because they reflect some of our most defining moments. In turn, this wider social network represents more resources that can be drawn upon to handle future difficulties.

In a culture that increasingly prizes positivity, we need opportunities to candidly express and experience pain and discomfort. One of the great paradoxes is that by being vulnerable with other people, sharing and disclosing painful events, we end up feeling more comfortable, connected, and courageous. Sometimes feeling bad is exactly what we need to live well. We might not become the greatest guitarist of all time, but learning how to sit with, work with, and channel our negative emotions can assuredly lead us to greater achievements, relationships, and a sense of happiness and meaning in life.

Todd B. Kashdan is professor of psychology and senior scientist at the Center for the Advancement of Well-Being at George Mason University. He served as the consultant on National Geographic Mind. *Robert Biswas-Diener is a researcher and professional trainer. This essay derives from their new book* The Upside of Your Dark Side: Why Being Your Whole Self—Not Just Your "Good" Self—Drives Success and Fulfillment.

FURTHER RESOURCES

BOOKS

Bray, Melissa A. *The Oxford Handbook of School Psychology*. New York: Oxford University Press, 2011.

Carter, Rita, and Christopher D. Frith. *Mapping the Mind*. Revised and updated ed. Berkeley: University of California Press, 2010.

Folkman, Susan. *The Oxford Handbook of Stress, Health, and Coping*. Oxford: Oxford University Press, 2011.

Frankl, Viktor E. *Man's Search for Meaning*. Boston: Beacon Press, 2006.

Hayes, Steven C. *Mindfulness and Acceptance: Expanding the Cognitive-Behavioral Tradition*. New York: Guilford Press, 2004.

Hunt, Morton M. *The Story of Psychology*. New York: Doubleday, 1993.

Kashdan, Todd B., and Robert Biswas-Diener. *The Upside of Your Down Side: Why Being Your Whole Self—Not Just Your "Good" Self—Drives Success and Fulfillment*. New York: Hudson Street Press, 2014.

Larsen, Randy J., and David M. Buss. *Personality Psychology: Domains of Knowledge about Human Nature*. 4th Ed. New York: McGraw-Hill, 2010.

Marcus, Gary F. *The Norton Psychology Reader*. New York: W. W. Norton, 2006.

Myers, David G. *Psychology*. 9th Ed. New York: Worth Publishers, 2010.

Nettles, Daniel. *Personality: What Makes You the Way You Are*. Oxford: Oxford University Press, 2007.

Snyder, C. R. *Oxford Handbook of Positive Psychology*. Oxford: Oxford University Press, 2009.

Sweeney, Michael S. Brain: *The Complete Mind*. Washington, D.C.: National Geographic Books, 2009.

———. *National Geographic Complete Guide to Brain Health*. Washington, D.C.: National Geographic Books, 2013.

Taylor, Jill Bolte. *My Stroke of Insight: A Brain Scientist's Personal Journey*. New York: Viking Penguin, 2008.

Valenstein, Elliot S. *Great and Desperate Cures: The Rise and Decline of Psychosurgery and Other Radical Treatments for Mental Illness*. New York: Basic Books, 1986.

ARTICLES

Acevedo, Bianca P., et al. "Neural Correlates of Long-Term Intense Romantic Love." *Social Cognitive and Affective Neuroscience* 7, no. 2 (2012): 145–59.

Alicke, M. D. "Global Self-Evaluation as Determined by the Desirability and Controllability of Trait Adjectives." *Journal of Personality and Social Psychology* 49, no. 6 (1985): 1621–30.

Baumeister, Roy F., and Mark R. Leary. "The Need to Belong: Desire for Interpersonal Attachments As a Fundamental Human Motivation." *Psychological Bulletin:* 497–529.

Behar, Evelyn, Ilyse Dobrow DiMarco, Eric B. Hekler, Jan Mohlman, and Alison M. Staples. "Current Theoretical Models of Generalized Anxiety Disorder (GAD): Conceptual Review and Treatment Implications." *Journal of Anxiety Disorders:* 1011–23.

Biswas-Diener, Robert, Todd B. Kashdan, and Gurpal Minhas. "A Dynamic Approach to Psychological Strength Development and Intervention." *Journal of Positive Psychology* (2011): 106–18.

Blackwell, Lisa S., Kali H. Trzesniewski, and Carol Sorich Dweck. "Implicit Theories of Intelligence Predict Achievement Across an Adolescent Transition: A Longitudinal Study and an Intervention." *Child Development:* 246–63.

Boeding, Sara E., Christine M. Paprocki, Donald H. Baucom, Jonathan S. Abramowitz, Michael G. Wheaton, Laura E. Fabricant, and Melanie S. Fischer. "Let Me Check That for You: Symptom Accommodation in Romantic Partners of Adults with Obsessive–Compulsive Disorder." *Behaviour Research and Therapy:* 316–22.

Braffman, Wayne, and Irving Kirsch. "Imaginative Suggestibility And Hypnotizability: An Empirical Analysis." *Journal of Personality and Social Psychology:* 578–87.

Bryan, Christopher J. et al. "Motivating Voter Turnout by Invoking the Self." *Proceedings of the National Academy of Sciences of the United States of America* 108, no. 31 (2011): 12653–56.

Buss, David. "Sex Differences in Human Mate Preferences: Evolutionary Hypotheses Tested in 37 Cultures." *Cambridge Journals Online:* 1–14.

Butler, A., J. Chapman, E. Forman, and A. Beck. "The Empirical Status of Cognitive-Behavioral Therapy: A Review of Meta-analyses." *Clinical Psychology Review:* 17–31.

Cacioppo, John T., et al. (2007) "Social Neuroscience: Progress and Implications for Mental Health." *Perspectives on Psychological Science* 2, no. 2: 99–123.

Cikara, Mina, Matthew M. Botvinick, and Susan T. Fiske. "Us Versus Them: Social Identity Shapes Neural Responses to Intergroup Competition and Harm." *Psychological Science* 22, no. 3 (2011).

Confer, J. C., J. A. Easton, D. S. Fleischman, C. D. Goetz, D. M. G. Lewis, C. Perilloux, and D. M. Buss. "Evolutionary Psychology: Controversies, Questions, Prospects, and Limitations." *American Psychologist* 65, no. 2 (2010): 110–126.

Duckworth, Angela L., Christopher Peterson, Michael D. Matthews, and Dennis R. Kelly. "Grit: Perseverance and Passion for Long-term Goals." *Journal of Personality and Social Psychology:* 1087–1101.

Dunham, Lena. "Difficult Girl." *The New Yorker,* Sept. 1, 2014.

Eisenberger, N. I. "Broken Hearts and Broken Bones: A Neural Perspective on the Similarities Between Social and Physical Pain." *Current Directions in Psychological Science:* 42–47.

Elliot, A. J., and T. M. Thrash. "Approach-avoidance motivation in personality: Approach and avoidance temperaments and goals." *Journal of Personality and Social Psychology* 82, no. 5 (2002): 804–18.

Emmons, Robert A. "Personal Strivings: An Approach to Personality and Subjective Well-being." *Journal of Personality and Social Psychology:* 1058–68.

Feng Yu et al. "A New Case of Complete Primary Cerebellar Agenesis: Clinical and Imaging Findings in a Living Patient." *Brain* (Aug. 2014).

Fincham, Frank D., Steven R. H. Beach, and Joanne Davila. "Longitudinal Relations Between Forgiveness and Conflict Resolution in Marriage." *Journal of Family Psychology:* 542–45.

Fromkin, Victoria, Stephen Krashen, Susan Curtiss, David Rigler, and Marilyn Rigler. "The Development Of Language In Genie: A Case Of Language Acquisition Beyond The 'Critical Period'." *Brain and Language* (1974): 81–107.

Gambino, Megan. "Are You Smarter Than Your Grandfather? Probably Not." Available online at www.smithsonianmag.com/science-nature/are-you-smarter-than-your-grandfather-probably-not-150402883/?all&no-ist.

Greengross, Gil, and Geoffrey F. Miller. "The Big Five Personality Traits of Professional Comedians Compared to Amateur Comedians, Comedy Writers, and College Students." *Personality and Individual Differences:* 79–83.

Guéguen, Nicolas, et al. "Men's Music Ability and Attractiveness to Women in a Real-Life Courtship Context." *Psychology of Music* 42, no. 4 (2014): 545–49.

Hunter, J. P., and M. Csikszentmihalyi. (2003). "The Positive Psychology of Interested Adolescents." *Journal of Youth and Adolescence* 32, no. 1: 27–35.

Karney, Benjamin R., and Thomas N. Bradbury. "Neuroticism, Marital Interaction, and the Trajectory of Marital Satisfaction." *Journal of Personality and Social Psychology:* 1075–92.

Kashdan, Todd B., and Jonathan Rottenberg. "Psychological Flexibility As a Fundamental Aspect of Health." *Clinical Psychology* Review: 865–78.

Kashdan, Todd B., William E. Breen, and Terri Julian. "Everyday Strivings in War Veterans with Posttraumatic Stress Disorder: Suffering From a Hyper-Focus on Avoidance and Emotion Regulation." *Behavior Therapy:* 350–63.

Keyes, Corey L. M. "Mental Illness and/or Mental Health? Investigating Axioms of the Complete State Model of Health." *Journal of Consulting and Clinical Psychology:* 539–48.

Knafo, Ariel, and Avi Assor. "Motivation for Agreement with Parental Values: Desirable When

Autonomous, Problematic When Controlled." *Motivation and Emotion* (2007): 232–45.

Lilienfeld, Scott O. "Why Ineffective Psychotherapies Appear to Work: A Taxonomy of Causes of Spurious Therapeutic Effectiveness." *Perspectives on Psychological Science* (1745-6916) 9, no. 4: 355.

Lilienfeld, Scott O., Lorie A. Ritschel, Steven Jay Lynn, Robin L. Cautin, and Robert D. Latzman. "Why Many Clinical Psychologists Are Resistant to Evidence-Based Practice: Root Causes and Constructive Remedies." *Clinical Psychology Review:* 883–900.

Longe, Olivia, Frances A. Maratos, Paul Gilbert, Gaynor Evans, Faye Volker, Helen Rockliff, and Gina Rippon. "Having a Word with Yourself: Neural Correlates of Self-Criticism and Self-Reassurance." *NeuroImage* 49, no. 2 (2010): 1849–56.

Lyubomirsky, S., King, L., & Diener, E. "The benefits of frequent positive affect: Does happiness lead to success?" *Psychological Bulletin* 131, no. 6 (2005): 803–55.

Macrae, C. N. "Medial Prefrontal Activity Predicts Memory for Self." *Cerebral Cortex:* 647–54.

McAdams, D. P., J. Reynolds, M. Lewis, A. H. Patten, and P. J. Bowman. "When Bad Things Turn Good and Good Things Turn Bad: Sequences of Redemption and Contamination in Life Narrative and Their Relation to Psychosocial Adaptation in Midlife Adults and in Students." *Personality and Social Psychology Bulletin:* 474–85.

McGregor, Ian, Mark P. Zanna, John G. Holmes, and Steven J. Spencer. "Compensatory Conviction in the Face of Personal Uncertainty: Going to Extremes and Being Oneself." *Journal of Personality and Social Psychology:* 472–88.

McNulty, J. K., and V. M. Russell. "When 'Negative' Behaviors Are Positive: A Contextual Analysis of the Long-Term Effects of Problem-Solving Behaviors on Changes in Relationship Satisfaction." *Journal of Personality and Social Psychology* 98, no. 4 (2010), 587.

Murphy, M. L. M., G. M. Slavich, N. Rohleder, and G. E. Miller. "Targeted Rejection Triggers Differential Pro- and Anti-Inflammatory Gene Expression in Adolescents as a Function of Social Status." *Clinical Psychological Science* (2012): 30–40.

Neff, Kristin D., Stephanie S. Rude, and Kristin L. Kirkpatrick. "An Examination of Self-Compassion in Relation to Positive Psychological Functioning and Personality Traits." *Journal of Research in Personality:* 908–16.

Ochsner, Kevin N., Silvia A. Bunge, James J. Gross, and John D. E. Gabrieli. "Rethinking Feelings: An fMRI Study of the Cognitive Regulation of Emotion." *Journal of Cognitive Neuroscience:* 1215–29.

Orne, M. T., and F. J. Evans. "Social Control in the Psychological Experiment: Antisocial Behavior and Hypnosis." *Journal of Personality and Social Psychology* 1, no. 3 (1965): 189–200.

Orser, B. A., Mazer, C. D., and A. J. Baker. "Awareness During Anesthesia." *CMAJ : Canadian Medical Association Journal* 178, no. 2 (2008): 185–88.

Owen, A. M. "Detecting Awareness in the Vegetative State." *Science:* 1402.

Paquette, V. "'Change the Mind and You Change the Brain': Effects of Cognitive-Behavioral Therapy on the Neural Correlates of Spider Phobia." *NeuroImage:* 401–09.

Petrosino, Anthony, Carolyn Turpin-Petrosino, and John Buehler. "Scared Straight and Other Juvenile Awareness Programs for Preventing Juvenile Delinquency: A Systematic Review of the Randomized Experimental Evidence." *Annals of the American Academy of Political and Social Science:* 41–62.

Piurko, Yuval, Shalom H. Schwartz, and Eldad Davidov. "Basic Personal Values and the Meaning of Left-Right Political Orientations in 20 Countries." *Political Psychology:* 537–61.

Postma, Albert, Gerry Jager, Roy P. C. Kessels, Hans P. F. Koppeschaar, and Jack Van Honk. "Sex Differences for Selective Forms of Spatial Memory." *Brain and Cognition:* 24–34.

Rahhal, T. A., C. P. May, and L. Hasher. "Truth and Character: Sources That Older Adults Can Remember." *Psychological Science* (2002): 101–5.

Raynor, Douglas A., and Heidi Levine. "Associations Between the Five-Factor Model of Personality and Health Behaviors Among College Students." *Journal of American College Health:* 73–82.

Reber, A. "Implicit Learning of Artificial Grammars." *Journal of Verbal Learning and Verbal Behavior:* 855–63.

Riskind, John H., and Nathan L. Williams. "The Looming Cognitive Style and Generalized Anxiety Disorder: Distinctive Danger Schemas and Cognitive Phenomenology." *Cognitive Therapy and Research:* 7–27.

Roese, N. J. "Sex Differences in Regret: All for Love or Some for Lust?" *Personality and Social Psychology Bulletin:* 770–80.

Ruminjo, Anne, and Boris Mekinulov. "A Case Report of Cotard's Syndrome." *Psychiatry* (Edgmont) 5, no. 6 (2008): 28–29.

Rusbult, Caryl E., Eli J. Finkel, and Madoka Kumashiro. "The Michelangelo Phenomenon." *Current Directions in Psychological Science:* 305–9.

Ryff, Carol D. "Psychological Well-Being in Adult Life." *Current Directions in Psychological Science:* 99–104.

Sacks, Oliver. "Face-Blind." *The New Yorker,* Aug. 30, 2010.

Schwartz, Shalom H. "An Overview of the Schwartz Theory of Basic Values." Online Readings in Psychology and Culture. Available online at http://scholarworks.gvsu.edu/cgi/viewcontentcgi?article=1116&context=orpc.

Shenk, Joshua. "What Makes Us Happy?" *The Atlantic,* June 1, 2009.

Shuwairi, S. M., M. K. Albert, and S. P. Johnson. "Discrimination of Possible and Impossible Objects in Early Infancy." *Journal of Vision* (2005): 528.

Steger, Michael F., and Todd B. Kashdan. "The Unbearable Lightness of Meaning: Well-Being and Unstable Meaning in Life." *The Journal of Positive Psychology:* 103–15.

Sternberg, Robert J., and David A. Kalmar. "When Will the Milk Spoil? Everyday Induction in Human Intelligence." *Intelligence:* 185–203.

Sternberg, Robert J., and Joyce Gastel. "If Dancers Ate Their Shoes: Inductive Reasoning with Factual and Counterfactual Premises." *Memory & Cognition:* 1–10.

Todman, D. "Inspiration from Dreams in Neuroscience Research." *The Internet Journal of Neurology* 9, no. 1.

Tranel, D., and A. Damasio. "Knowledge Without Awareness: An Autonomic Index of Facial Recognition by Prosopagnosics." *Science:* 1453–54.

Trut, Lyudmila N. "Early Canid Domestication: The Farm-Fox Experiment." *American Scientist* 87, no. 2: 160–69.

Tsay, Cynthia J. "Julius Wagner-Jauregg and the Legacy of Malarial Therapy for the Treatment of General Paresis of the Insane." *The Yale Journal of Biology and Medicine* 86, no. 2 (2013): 245–54.

Tuckman, Bruce. "Developmental Sequence in Small Groups." *Psychological Bulletin:* 384–99.

Uskul, A. K., and H. Over. "Responses to Social Exclusion in Cultural Context: Evidence from Farming and Herding Communities." *Journal of Personality and Social Psychology* 106, no. 5 (2014): 752–71.

Walton, G. M., and G. L. Cohen. "A Brief Social-Belonging Intervention Improves Academic and Health Outcomes of Minority Students." *Science:* 1447–51.

Walton, Gregory M. "The New Science of Wise Psychological Interventions." *Current Directions in Psychological Science* 23, no. 1 (2014): 73–82.

Westen, Drew. "The Scientific Legacy of Sigmund Freud: Toward a Psychodynamically Informed Psychological Science." *Psychological Bulletin:* 333–71.

Winkielman, Piotr, and Kent C. Berridge. "Unconscious Emotion." *Current Directions in Psychological Science:* 120–3.

Wyland, Carrie L., William M. Kelley, C. Neil Macrae, Heather L. Gordon, and Todd F. Heatherton. "Neural Correlates of Thought Suppression." *Neuropsychologia:* 1863–67.

Wynn, Karen. "Addition and Subtraction by Human Infants." *Nature:* 749–50.

WEBSITES

APA Policy Statement on Evidence-Based Practice in Psychology www.apa.org/practice/guidelines/evidence-based-statement.aspx?item=5

Cognitive Behavioral Therapy www2.nami.org/Content/NavigationMenu/Inform_Yourself/About_Mental_Illness/About_Treatments_and_Supports/Cognitive_Behavioral_Therapy1.htm

Mental Illness Facts and Numbers www2.nami.org/factsheets/mentalillness_factsheet.pdf

ABOUT THE AUTHOR

PATRICIA DANIELS is a writer and editor with a particular interest in science and history. Among her books for National Geographic are *The Body: A Complete User's Guide, The New Solar System, Great Empires,* and *The National Geographic Almanac of World History.* She is also a member of the board of editors for Macmillan's *Discoveries in Modern Science.* She has been a managing editor for Time-Life Books and a senior writer for *National Wildlife* magazine. She lives in State College, Pennsylvania, with her husband, a college professor.

ACKNOWLEDGMENTS

I couldn't have written this book without help from many smart and talented people. In particular, I'd like to thank Dr. Todd B. Kashdan and his colleagues Fallon Goodman and Kevin Young for their expert guidance and extensive research. I'm also grateful, as always, for the editorial direction and wisdom of editors Susan Tyler Hitchcock and Barbara Payne; for Elisa Gibson's expert design; and for Michelle Cassidy's research and writing skills. And last but not least, many thanks to my husband, Jim Tybout, for listening to endless mind-related anecdotes at the dinner table.`

(see page 235)

How to score your assessment scales

After completing your striving assessment scales, code your answers as approach-themed versus avoidance-themed (see below). When it comes to the pursuit of well-being, it is better to have more approach strivings than avoidance strivings. After coding, if you want to work on improving your quality of life, consider rewording or reframing your avoidance strivings into approach language. You can also see for yourself what you are spending your finite currency of time and effort on and what rewards you are getting.

APPROACH VS. AVOIDANCE

Does the striving refer to something positive or negative?

Does the person wish to approach, obtain, achieve, or keep the object of the striving or do they wish to avoid, prevent, or get rid of the object of the striving?

Is the person trying "not to do something"? (Usually the words "avoid," "not," or "don't" will give it away.)

(if not clearly avoidant, then code as approach):

"Not to feel bad when people dislike me for no known reason"

"Avoid escapism (fantasy and speculation about future)"

"Avoid letting anything upset me"

"Not to be possessive with my boyfriend"

"Not to feel inferior in social gatherings"

"Don't procrastinate"

"Restrain from arguing"

"Smoke less, drink less"

ABOUT THE CONSULTANTS

A world-recognized authority on well-being, strengths, social relationships, stress, and anxiety, TODD B. KASHDAN has published more than 150 scholarly articles and is the author or coauthor of several books, including *The Upside of Your Dark Side: Why Being Your Whole Self—Not Just Your "Good" Self—Drives Success and Fulfillment.* Kashdan is professor of psychology and senior scientist at the Center for the Advancement of Well-Being at George Mason University. Honors include early career awards from the American Psychological Association, Association for Behavioral and Cognitive Therapies, and International Society for Quality of Life Studies. He's a twin with twin eight-year-old daughters, and has plans to populate the world with great conversationalists.

KEVIN YOUNG is a doctoral student in the clinical psychology program at George Mason University. His research interests include the exploration of methods to maximize personal and interpersonal well-being, including the identification and application of character strengths. Young also consults with international academic and professional organizations regarding the measurement of staff and student well-being, as well as the development and implementation of positive psychology interventions.

FALLON GOODMAN is a doctoral student in clinical psychology and research fellow at the Center for the Advancement of Well-Being at George Mason University. Goodman's scholarly interests include well-being measurement and intervention, social anxiety, and self-regulation. She is passionate about conducting and disseminating research that can be used to improve people's lives.

ILLUSTRATIONS CREDITS

Front cover: (twins) Martin Schoeller; (flowers) Emesilva/iStockphoto; (yoga pose) Vinogradov Illya/Shutterstock; (Freud) Hulton-Deutsch Collection/Corbis UK Ltd; (DNA structure) Creations/Shutterstock; (equations) Gill Button; (brain) Sebastian Kaulitzki/Shutterstock; (baby) Victoria Penafiel/Getty Images; (people forming heart) LWA/Dann Tardif/Getty Images.

Back cover: (couple) Matthew Nigel/Shutterstock; (footprints in sand) Alta Oosthuizen/Shutterstock; (sleep study) Maggie Steber; (man with roots) Andrea Danti/Shutterstock; (ringed hand) Maria Ramos Urbano/Getty Images; (children with masks) Image Source/Getty Images; (brain activity) Laguna Design/Science Photo Library.

2-3 Jimmy Anderson/Getty Images; 4 Mikey Schaefer; 7 Bertrand Demee/Getty Images; 8 Dwight Smith/Shutterstock; 12 Matt Propert/National Geographic Image Collection; 14 De Visu/Shutterstock; 15 Robert Leighton The New Yorker Collection/The Cartoon Bank/The Condé Nast Publications Ltd; 17 Darren Ashcroft; 18 Analytic couch in Sigmund Freud's study (photo)/Freud Museum, London, UK/Bridgeman Art Library; 19 Corbis UK Ltd; 20-21 bikeriderlondon/Shutterstock; 22 Vincent J. Musi; 23 (UPRT) jmcdermottillo/Shutterstock; 23 (LORT) Serg64/Shutterstock; 24 Blue Lantern Studio/Corbis UK Ltd; 25 Ariel Skelley/Blend Images/Corbis UK Ltd; 26 LWA/Dann Tardiff/Getty Images; 28 andresr/iStockphoto; 29 stevanovicigor/Deposit Photos; 30 yaruta/iStockphoto; 31 Gyvafoto/Shutterstock; 33 Official White House Photo by Pete Souza/White House; 34 Jiri Hera/Shutterstock; 35 rollover/iStockphoto; 36 David Malan/Getty Images; 37 Brand New Images/Getty Images; 38 Evgeny Atamanenko/Shutterstock; 40 Mehau Kulyk/SPL/Getty Images; 41 Fernando Da Cunha/BSIP/Corbis UK Ltd; 42 (UP) Gill Button; 42 (LO) maniaroom/Shutterstock; 43 Jeff Vanuga/Corbis UK Ltd; 44 yulkapopkova/iStockphoto; 45 Davis Meltzer/National Geographic Image Collection; 46 CGinspiration/iStockphoto; 47 RTimages/iStockphoto; 48 Daily Herald Archive/SSPL/Getty Images; 49 Mike Baldwin/www.CartoonStock.com; 50 Eric Isselee/Shutterstock; 51 Katia Platonova & Andrey Kazakov/Photostudio; 52 shapecharge/iStockphoto; 53 @erics/Shutterstock; 54 Maggie Steber; 55 PM Images/Getty Images; 56 ImageZoo/Alamy; 57 SensorSpot/Getty Images; 58 (UP) Beau Lark/Corbis UK Ltd; 58 (LO) Joe Belanger/123RF; 59 VasjaKoman/iStockphoto; 60 Jannoon028/iStockphoto; 61 Radius Images/Corbis UK Ltd; 62 Luis Louro/Shutterstock; 63 SpeedKingz/Shutterstock; 64 (LE) LiliGraphie/Shutterstock; 64 (RT) Elzbieta Sekowska/Shutterstock; 65 aletia/Deposit Photos; 66 stockillustration/Shutterstock; 67 ArtFamily/Shutterstock; 68 Brian Snyder/Reuters/Corbis UK Ltd; 69 skeeg/Getty Images; 71 Kaponia Aliaksei/Shutterstock; 72 bikeriderlondon/Shutterstock; 74 Classic Image/Alamy; 75 GlobalP/iStockphoto; 76 Gary Brown/Science Photo Library; 77 Sam Falk/Science Photo Library; 78 Image Source/Getty Images; 80 (LE) opicobello/Shutterstock; 80 (RT) Shelepov Stanislav/Shutterstock; 81 Science Photo Library/Alamy; 82 Margot Hartford/Alamy; 83 Inara Prusakova/Shutterstock; 84 Lagui/Shutterstock; 85 Lemon Tree Images/Shutterstock; 86 Grischa Georgiew/Shutterstock; 87 Andrea Danti/Shutterstock; 88 David McLain/Getty Images; 89 narokzaad/iStockphoto; 90 Image Source/Getty Images; 91 Maydaymayday/Getty Images; 92 GlobalP/iStockphoto; 93 xPACIFICA/National Geographic/Corbis UK Ltd; 94 Hagen/www.CartoonStock.com; 95 BraunS/iStockphoto; 96 Johnny Greig/Getty Images; 97 Associated Press Contributors/Press Association Images; 98 Lightspring/Shutterstock; 99 benchart/Shutterstock; 100 Associated Press/Press Association Images; 101 Kelly-Mooney Photography/Corb/Corbis UK Ltd; 102 ArtMarie/iStockphoto; 103 Roy Scott/Ikon Images/Corbis UK Ltd; 104 Kazlova Iryna/Shutterstock; 105 BLOOM Image/Getty Images; 108 Henrik Sorensen/Getty Images; 110 Picsfive/Shutterstock; 111 FineArt/Alamy; 112 James Francis/Shutterstock; 113 Mandy Godbehear/Shutterstock; 115 vita khorzhevska/Shutterstock; 116 Gandee Vasan/Getty Images; 117 Deshakalyan Chowdhury/Getty Images; 118 Baloo/www.CartoonStock.com; 119 Bettmann/Corbis UK

Ltd; 120 DEA/G. Nimatallah/Getty Images; 121 dutchkris/iStockphoto; 122 Frans Lanting Photography; 123 hermione13/123RF; 124 Stephen Rees/Shutterstock; 125 Alberto Ruggieri/Illustration Works/Corbis UK Ltd; 126 Hagen/ www.CartoonStock.com; 127 Photographic Art Viet Nam/Shutterstock; 129 Robert Churchill/ iStockphoto; 130-131 Jules_Kitano/ Shutterstock; 132 -nelis-/iStockphoto; 133 Associated Press/Press Association Images; 134 marekuliasz/Shutterstock; 135 Titova E/Shutterstock; 136 Mitchell Funk/Getty Images; 137 altrendo images/Getty Images; 138 Mike Kemp/Blend Images/Corbis UK Ltd; 140 Bettmann/Corbis UK Ltd; 141 SoRad/Shutterstock; 142 Randy Faris/Corbis UK Ltd; 143 Milles Studio/Shutterstock; 145 (CT) Ljupco Smokovski/Shutterstock; 145 (RT) Chatchawan/Shutterstock; 146 Blue Line Pictures/ Getty Images; 148 David Crockett/ Shutterstock; 149 Radius Images/ Corbis UK Ltd; 150 raywoo/ Deposit Photos; 151 Ernesto Víctor Saúl Herrera Hernández/iStockphoto; 152 Stephen Mcsweeny/ Shutterstock; 153 lorenzo rossi/ Alamy; 154 (UP) Walter Daran/ Getty Images; 154 (LO) Ljupco Smokovski/Shutterstock; 155 Ellis Nadler/www.CartoonStock .com; 156 tovovan/Shutterstock; 157 Blasius Erlinger/Corbis UK Ltd; 158 Hero Images/Getty Images; 159 PathDoc/Shutterstock; 160 geopaul/iStockphoto; 161 Sergey Nivens/ Shutterstock; 162 Ho Yeow Hui/ Shutterstock; 163 Carolyn Iagattuta/Getty Images; 164 Vincent van Gogh/Getty Images; 165 DanielW/ Shutterstock; 166 alastar89/Shutterstock; 167 blueenayim/iStockphoto; 168 Cory Richards/National Geographic Creative; 170 John Lund/Blend Images/Corbis UK Ltd; 172 inithings/Shutterstock; 173 Anna Maltseva/Shutterstock; 174 (LE) Netfalls—Remy Musser/Shutterstock; 174 (RT) aleksandr hunta/ Shutterstock; 175 Image Source/ Getty Images; 176 lassedesignen/ Shutterstock; 177 Richard Peterson/ Shutterstock; 178 Diego Cervo/ Shutterstock; 179 jmcdermottillo/ Shutterstock; 180 Jodi Cobb; 181 akova /iStockphoto; 182 Dave Nagel/Getty Images; 183 Pressmaster/Shutterstock; 184 Dean Mitchell/iStockphoto; 185 joebelanger/ iStockphoto; 186 marekuliasz/Shutterstock; 188 amriphoto/iStockphoto; 189 Randy Glasbergen; 190 choreograph/123RF; 193 (LE) Savany/iStockphoto; 193 (RT) Guzel Studio/Shutterstock; 194 Randy Glasbergen; 195 Michelle D. Milliman/Shutterstock; 196 Andrew Innerarity/Reuters/Corbis UK Ltd; 197 Chip Somodevilla/Getty Images; 198 Nicolesa/Shutterstock; 199 Bruce Bennett/Getty Images; 202 Elena Kulikova/Getty Images; 205 Brain light/Alamy; 206 ilyast/Getty Images; 207 Brian Jackson/Alamy; 208 UpperCut Images/ Alamy; 209 Jessica Peterson/Getty Images; 210 aetmeister/iStockphoto; 211 Illustration Works/ Alamy; 213 Jose Luis Pelaez Inc/ Getty Images; 214 Serebryakov Andrei/ITAR-TASS/Corbis UK Ltd; 215 PathDoc/Shutterstock; 216 Alexstar/Deposit Photos; 217 Garry Gay/Alamy; 218 Robin Skjoldborg/ Getty Images; 219 Robert Red/Shutterstock; 220 Photo by wili_hybrid/ Ville Miettinen; 221 Mark Anderson; 222 4zevar/Shutterstock; 223 Tony Hutchings/Getty Images; 225 Ira Block Photography; 226 MTMCOINS/iStockphoto; 228 Peter Dazeley/Getty Images; 229 Patrick Strattner/Getty Images; 230 Daniel Allan/Cultura/Corbis UK Ltd; 231 Yaroslav Seheda/Solent News/Rex Features; 232 Eugene Novikov/Shutterstock; 234 (LO) mattjeacock/ iStockphoto; 234 (UP) Thomas Barwick/Getty Images; 236 Michael Krinke/iStockphoto; 237 Bettmann/ Corbis UK Ltd; 238 Blulz60/Shutterstock; 239 JGI/Jamie Grill/Getty Images; 240 Picsfive/Shutterstock; 241 Randy Lincks/Corbis UK Ltd; 242 Marina Ramos Urbano/ Getty Images; 243 Jason Hosking/ Corbis UK Ltd; 244 (LO) lsantilli/ Shutterstock 244 (UP) Hugh Sitton/ Corbis UK Ltd; 245 Eye-Stock/ Alamy; 246 David Lee/Shutterstock; 247 All Over Sweden/Splash News/Corbis UK Ltd; 248 William Berry/Shutterstock; 249 Subhadeep Mukherjee; 250 lamnee/Shutterstock; 251 hcchoo/Getty Images; 253 Kirklandphotos/Getty Images; 254 William Andrew/Getty Images; 256 (LE) Universal History Archive/Universal Images /Rex Features; 256 (RT) Universal History Archive/Universal Images/Rex Features; 257 DesignPrax/Shutterstock; 258 bookzaa/Shutterstock; 259 Gahan Wilson The New Yorker Collection/ The Cartoon Bank/The Condé Nast Publications Ltd; 260 Imagezoo/Corbis UK Ltd; 261 enviromantic/iStockphoto; 263 Ljm Photo/Design Pics/ Corbis UK Ltd; 264 Sean Gallagher/ National Geographic Creative/Corbis UK Ltd; 266 lushik/Getty Images; 267 Ryan McVay/Getty Images; 268 Superstock/Corbis UK Ltd; 269 johnkworks/Shutterstock; 270 Image Source/Getty Images; 271 Chaiwat Subprasom/ Reuters/Corbis UK Ltd; 273 Photawa/iStockphoto; 274 Jiri Flogel/ Shutterstock; 275 Gary Bass/Shutterstock; 276 Graham Denholm/Getty Images; 277 mediaphotos/Getty Images; 278 Cassandra Hannagan/ Corbis UK Ltd; 280 Jeremy Woodhouse/Getty Images; 281 Ji í Truba /123RF; 282 Kamala Kannan; 283 Diane Cook and Len Jenshel.

INDEX

Page numbers in **bold** indicate illustrations and non-text material.

J

K

L

M

R

S

T

U

V

W

Y

NATIONAL GEOGRAPHIC MIND
PATRICIA DANIELS

The National Geographic Society is one of the world's largest nonprofit scientific and educational organizations. Founded in 1888 to "increase and diffuse geographic knowledge," the member-supported Society works to inspire people to care about the planet. Through its online community, members can get closer to explorers and photographers, connect with other members around the world, and help make a difference. National Geographic reflects the world through its magazines, television programs, films, music and radio, books, DVDs, maps, exhibitions, live events, school publishing programs, interactive media, and merchandise. *National Geographic* magazine, the Society's official journal, published in English and 38 local-language editions, is read by more than 60 million people each month. The National Geographic Channel reaches 440 million households in 171 countries in 38 languages. National Geographic Digital Media receives more than 25 million visitors a month. National Geographic has funded more than 10,000 scientific research, conservation, and exploration projects and supports an education program promoting geography literacy. For more information, visit www.nationalgeographic.com.

For more information, please call 1-800-NGS LINE (647-5463) or write to the following address:

National Geographic Society
1145 17th Street NW
Washington, D.C. 20036-4688 U.S.A.

For information about special discounts for bulk purchases, please contact National Geographic Books Special Sales: ngspecsales@ngs.org

For rights or permissions inquiries, please contact National Geographic Books Subsidiary Rights: ngbookrights@ngs.org

Library of Congress Cataloging-in-Publication Data

Daniels, Patricia, 1955-
National Geographic mind : a scientific guide to who you are, how you got that way, and how to make the most of it / Patricia Daniels ; foreword by Todd B. Kashdan.
pages cm
Includes bibliographical references and index.
ISBN 978-1-4262-1672-5 (hardcover : alk. paper) -- ISBN 978-1-4262-1673-2 (hardcover deluxe edition : alk. paper)
1. Psychology. 2. Developmental psychology. 3. Human behavior. I. Title.
BF145.D36 2016
155.2--dc23

2015013691

Printed in the United States of America

15/QGT-CML/1